A-level Study Guide

Chemistry

Philip Barratt

Michael Cox

Series Consultants: Geoff Black and Stuart Wall
Project Manager: Gillian Ragsdale

Pearson Education Limited
Edinburgh Gate, Harlow
Essex CM20 2JE, England
and Associated Companies throughout the world

© Pearson Education Limited 2000

First published 2000

British Library Cataloguing in Publication Data
A catalogue entry for this title is available from the British Library.

ISBN 0-582-43176-X

Set by 35 in Univers, Cheltenham
Printed by Ashford Colour Press, Gosport, Hants

Atoms, ions and molecules

You should already know that everything is made of atoms. For A-level, you will need to know what atoms are made of and how they form ions, simple molecules and giant structures. You will also need to understand how the properties of elements and compounds depend upon their structures and the forces that hold them together.

Exam themes

→ Write and balance different types of chemical equation and work out empirical formulae using masses or percentage composition by mass

→ Work out numbers of electrons, protons and neutrons using atomic and mass number and give electron configurations using s, p and d notation and the arrows in boxes method

→ Use mass spectra to work out A_r values

→ Use the Avogadro constant and do calculations on reacting masses, gas volumes and concentrations of solutions

→ Interpret plots of 1st ionization energy against atomic number

→ Draw dot and cross diagrams to show ionic and covalent bonding; draw examples and describe ionic lattices, simple molecules, giant covalent structures and metallic bonding. Relate the properties of a substance to the type of bonding it contains

→ Describe the main types of intermolecular forces between simple molecules

→ Work out shapes of molecules using VSEPR theory and work out whether molecules are polar

→ Use $pV = nRT$ and the gas laws

Topic checklist

O AS ● A2	AQA	CCEA	EDEXCEL	OCR	WJEC
Formulae and equations	O	O	O	O	O
Atoms and isotopes	O	O	O	O	O
Amount of substance and the mole	O	O	O	O	O
Atomic structure	O	O	O	O	O
Structure and bonding in elements	O	O	O	O	O
Structure and bonding in covalent compounds	O	O	O	O	O
Structure and bonding in ionic compounds	O	O	O	O	O
Intermolecular forces	O	O	O	O	O
Gases	O	O	O	O	O

Formulae and equations

Every element has a unique symbol. Chemists combine symbols to write formulae of chemicals and use formulae to write equations for chemical reactions. A formula can be shorthand for the name, composition and structure of a substance. A formula can represent an amount of substance and an equation can be the basis of chemical calculations involving masses, volumes, concentrations, etc.

Types of formulae

Empirical (simplest) formula

→ shows *ratios of the numbers* of each atom (or ion) in a compound
→ shows the *amounts of each element* in one formula mass

You should know how to use the composition by mass to find the empirical formula of a compound.

1 Divide each mass (or % mass) by the A_r of each element.
2 Calculate the simplest whole number ratios.

Example: An oxide of rubidium

element	% mass ÷ A_r	=	mole ratio
Rb	84.2 ÷ 85.5 = 0.99		1
O	15.8 ÷ 16.0 = 0.99		1

The empirical formula is RbO.

Ionic formula

The ionic formula of a binary ionic compound is usually but not always the same as the empirical formula. The oxide of rubidium in the above example has a formula mass of 203 g mol^{-1} and its ionic formula is $(Rb^+)_2O^{2-}$. Ionic compounds form crystals containing millions of ions arranged in a regular structure based on a **unit cell**.
Ionic formulae are related to

→ the **radius ratios** of the ions in a unit cell
→ the **coordination numbers** of the ions in the crystal

If you know the formulae of the constituent ions you can deduce the formula of an ionic compound by making the charges balance.

Example: Magnesium bromide
Mg^{2+} is two positive and Br^- is one negative, so one magnesium cation needs two bromide anions to balance the charges: $Mg^{2+}(Br^-)_2$.

Molecular formula

This is for elements and compounds that form **covalent molecules**. Molecular formulae show the *numbers of atoms* in molecules. The noble gases are monatomic and ozone is triatomic but the other elemental gases like chlorine, nitrogen and oxygen are diatomic. You can use the **octet rule** to predict the molecular formula of many simple compounds of the s- and p-block but *not* the d-block elements.

Example: Chlorine oxide

Chlorine is in group VII of the periodic table: $8 - 7 = 1$
Oxygen is in group VI of the periodic table: $8 - 6 = 2$

So one O atom needs two Cl atoms to balance: Cl_2O.

Structural formula

A structural formula shows the *arrangement* of atoms and groups in a molecule. Structural formulae are essential for organic compounds.

Types of equations ●●●

For **organic** reactions we often write unbalanced 'equations' showing only the structural formulae of the principal organic reactant(s) and product(s):

$$CH_3CH_2OH \xrightarrow{} CH_3CHO$$
$$\text{ethanol} \quad \text{oxidized to} \quad \text{ethanal}$$

For **inorganic** reactions we usually write balanced equations showing all the reactants and products.

Always write *balanced* equations for calculations.

Ordinary equation

The equation for the reaction of magnesium metal with dilute aqueous sulphuric acid is

$$Mg(s) + H_2SO_4(aq) \rightarrow MgSO_4(aq) + H_2(g)$$

Ionic equation

The ionic equation for the reaction is

$$Mg(s) + 2H^+(aq) \rightarrow Mg^{2+}(aq) + H_2(g)$$

because the aqueous sulphate ion $SO_4^{2-}(aq)$ is a **spectator ion**.

Ion–electron half-equations

We can write the above ionic equation in the form of two ion–electron half-equations.

$$Mg(s) \rightarrow Mg^{2+}(aq) + 2e^- \qquad \text{oxidation – loss of electrons}$$
$$2H^+(aq) + 2e^- \rightarrow H_2(g) \qquad \text{reduction – gain of electrons}$$

You can use **oxidation numbers** to help balance equations for redox reactions.

Checkpoint 4

Predict the molecular formula of the simplest compound formed by each of the following pairs of elements.
(a) chlorine and iodine
(b) sulphur and chlorine
(c) carbon and oxygen
(d) nitrogen and iodine

Examiner's secrets

Most marks are given for the correct formulae and balancing but it's a good idea to include the state symbols (s), (aq), (g) and (l) used properly.

Links

See page 74: redox reactions.

Exam question (2 min) answer: page 23

Write balanced ordinary and ionic equations for the reaction of

(a) $KOH(aq)$ with $HNO_3(aq)$

(b) $Ba(OH)_2(aq)$ with $HCl(aq)$

(Hint: these are strong acids and bases.)

Watch out!

In ionic equations and ion–electron half-equations the symbols *and* the charges must balance!

Atoms and isotopes

Experiments support the theory that atoms consist of a tiny nucleus of protons and neutrons surrounded by a large volume of space which, apart from the electrons, is empty. The negative charge of the electrons balances the positive charge of the protons. Most of the mass of an atom is in the nucleus. Isotopes of an element have different numbers of neutrons in their nuclei. Some isotopes are radioactive.

Electrons, protons and neutrons

The following typical diagram is not to scale because the nucleus is too small compared to the surrounding space occupied by the electrons.

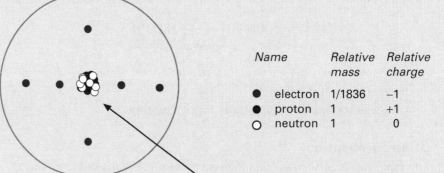

Name		Relative mass	Relative charge
●	electron	1/1836	−1
●	proton	1	+1
○	neutron	1	0

nucleus (not to scale): 6 protons and 7 neutrons

→ **Atomic number** (Z) is the number of protons in the nucleus of an atom.
→ **Mass number** (A) is the number of protons and neutrons in a nucleus.

Isotopes

All the atoms of a particular element must have the *same number* of protons in the nucleus but the number of neutrons may be *different*.

→ Isotopes are atoms with the *same* atomic number (Z) but *different* mass numbers (A).

Bromine has two naturally occurring isotopes: $^{79}_{35}\text{Br}$ and $^{81}_{35}\text{Br}$.

Radioactive isotopes

Some isotopes are unstable. Their nuclei spontaneously disintegrate and radiate *either* helium nuclei (alpha (α) particles) *or* electrons (beta (β) particles) but never both. Gamma (γ) rays (very high energy radiation) may also be given off. The **half-life** of a radioisotope is

→ the time taken for its radioactivity to fall to half of its initial value
→ independent of the mass of the radioisotope, so the radioactive decay of an isotope is a **first order rate of reaction**
→ characteristic of each radioisotope and unaffected by catalysts or changes in temperature

Checkpoint 1

What is the name and symbol of the element with this atomic structure?

Checkpoint 2

What is the atomic number and mass number of the atom represented in this diagram?

The jargon

Z is used to number the element's place in the periodic table. Nowadays Z is often called the proton number.

Checkpoint 3

How many neutrons are in the nucleus of each bromine isotope?

Checkpoint 4

(a) How many protons and neutrons are in a helium nucleus?
(b) What happens to Z and A when a nucleus emits
　(i) one α-particle
　(ii) one β-particle?

Stable isotopes

The atoms of the majority of elements exist as a mixture of isotopes. Most stable nuclides have an even number of protons and/or neutrons in the nucleus. For atomic numbers up to 20, the ratio of the number of neutrons to the number of protons in stable nuclides is $1:1$.

Relative atomic mass A_r ●●●

The actual mass of an atom is so small (the heaviest is only about 4×10^{-25} kg) that we use relative atomic masses. We compare the mass of atoms to the mass of carbon atoms.

→ The **relative atomic mass** (A_r) is the *ratio* of the *average mass* per atom of the natural isotopic composition of an element *to one-twelfth of the mass of an atom of nuclide* ^{12}C.

Abundance of stable isotopes

The natural isotopic abundance of elements has been very accurately determined from mass spectra obtained from **mass spectrometers**. You could be asked to

→ read these abundances from a given mass spectrum
→ use them to calculate the relative atomic mass

Example: Mass spectrum of magnesium

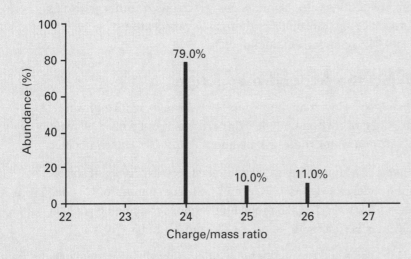

A_r(Mg) is $(79.0/100 \times 24 + 10.0/100 \times 25 + 11.0/100 \times 26) = 24.32$

The jargon

A *nuclide* is an atom of a specified mass number, A. All stable naturally occurring aluminium atoms exist as the nuclide $^{27}_{13}$Al.

Checkpoint 5

Why is the relative atomic mass of carbon itself 12.01 and not 12.00?

Links

See page 160: mass spectroscopy.

Watch out!

A_r(Mg) is sometimes called the atomic weight of magnesium but don't forget that A_r stands for a ratio, so 24.32 is a pure number – no units!

Examiner's secrets

Do not learn the A_r values. In an exam they are given to you to the first (or sometimes second) decimal place: see the periodic table on page 84.

Watch out!

Candidates often lose marks by taking the Z value instead of the A_r value from the periodic table.

Exam question (5 min) answer: page 23

35 and 37 are the mass numbers of the two stable naturally occurring isotopes of chlorine.

(a) State for each nuclide (i) the number of protons in the nucleus and
    (ii) the number of neutrons in the nucleus.

(b) Calculate the relative abundance of the two isotopes of chlorine (refer to the periodic table on page 84).

Amount of substance and the mole

When we read an equation we can see from the numbers and formulae how much of each reactant and product there is. The chemist's unit of 'how much' is the mole. You need to understand how it is defined and how we use it in calculations.

What is a mole?

One mole is the mass, in grams, of a substance as indicated by its formula. How can you find the mass of a mole? Look up the formula and **molar mass** in data books: e.g.

Formula	Na	Cl^-	H_2O	Na^+Cl^-
Molar mass/g mol^{-1}	23.0	35.5	18.02	58.5

You can also use the formula and the A_r values of the constituent elements to calculate the **relative formula mass**. Here are the steps:

1 The formula of water is H_2O.
2 $A_r(H) = 1.01$ and $A_r(O) = 16.0$.
3 The relative formula mass is $(2 \times 1.01 + 16.0) = 18.02$.
4 So the mass of one mole of H_2O would be 18.02 g.

What is an exact definition of a mole?

→ One mole (mol) is the amount of substance (n) that contains as many entities (atoms, moleces, electrons or other particles) specified by a formula as there are carbon atoms in 0.012 kg of carbon nuclide ^{12}C.

Using the Avogadro constant

Example 1: How many molecules are there in 18.02 g of water?
In 0.012 kg of carbon nuclide ^{12}C there are about 6.02×10^{23} atoms. So in 18.02 g of water there will be about 6.02×10^{23} H_2O molecules.

Example 2: How many molecules are there in 45.05 g of water?
A molar mass of H_2O is 18.02 g mol^{-1}. So, the amount of H_2O in 45.05 g water is 45.02/18.02 = 2.5 mol, and therefore the number of H_2O molecules is $(6.02 \times 10^{23} mol^{-1}) \times (2.5 mol) = 1.505 \times 10^{24}$.

How do we find the number of entities in any given amount of substance?

number of entities = $6.02 \times 10^{23} \times$ amount of substance
number of entities = $L \times$ amount of substance

→ The Avogadro constant is the proportionality constant connecting the number of entities (specified by a formula) to the amount of substance (expressed in moles).

Molar volume

This is the volume of one mole of substance under *specified conditions of temperature and pressure*.

In 1811 the Italian chemist Amadeo **Avogadro** stated this important principle:

The jargon

$n = m/M$
where n is the amount of substance in moles, m is the mass of substance in grams and M is the molar mass in grams per mole.

Checkpoint 1

Use the periodic table on page 84 to determine
(a) the mass in grams of
    (i)   1 mol of hydrogen atoms
    (ii) 0.5 mol nitrogen molecules N_2
    (iii) 1 mol ozone molecules O_3
(b) the molar mass of nitric acid, HNO_3

The jargon

The symbol for the Avogadro constant is an italic capital L.

Don't forget!

Since L is a number per mole, the units of L will be just mol^{-1}.

Examiner's secrets

You are always given the value of L to two (or three) significant figures. You do not need to remember it.

→ Equal volumes of gases at the *same temperature and pressure* contain the same number of molecules.

One mole of any gas occupies approximately 24 dm^3 at room temperature and 22.4 dm^3 at standard temperature and pressure (s.t.p.). According to the rule proposed by the French chemist Louis J. **Gay-Lussac** in 1809

→ when measured at the same temperature and pressure the volumes of gaseous reactants and products of a reaction will be in a simple ratio.

Example:
$$H_2(g) + Cl_2(g) \rightarrow 2HCl(g)$$
1 mol or 24 dm^3 1 mol or 24 dm^3 2 mol or 48 dm^3

so the gases are in the ratio 1 : 1 : 2.

Concentration ●●●

When 6.30 g of nitric acid is dissolved in water to produce exactly 1.00 dm^3 of solution, the resulting aqueous acid should be described as 'aqueous nitric, $HNO_3(aq)$, of concentration 0.100 mol dm^{-3}'.

→ Concentration is the amount of substance per unit volume of *solution* (not solvent) expressed in mol dm^{-3}.

Chemical calculations ●●●

You need to be able to use amounts of substances, molar gas volumes and concentrations of solutions to do calculations and solve problems based on balanced chemical equations. Here are two examples.

1. What volume of hydrogen at s.t.p. would be produced by reacting completely 0.486 g of magnesium with excess aqueous sulphuric acid? The equation for the reaction is

$$Mg(s) + H_2SO_4(aq) \rightarrow MgSO_4(aq) + H_2(g)$$

1 mol = 24.3 g would produce 1 mol = 22.4 dm^3 hydrogen so 0.486 g magnesium would produce $22.4 \times 0.486/24.3 = 0.448$ dm^{-3}.

2. What is the *minimum* volume of hydrochloric acid of concentration 0.01 mol dm^{-3} needed to react completely with 0.243 g of magnesium? The equation is $Mg(s) + 2HCl(aq) \rightarrow MgCl_2(aq) + H_2(g)$, so 0.243 g = 0.01 mol Mg needs 0.02 mol HCl(aq) or 2 dm^3 of solution.

Exam question (6 min)   answer: page 23

Calcium carbonate, $CaCO_3$, is obtained from limestone rock by open quarrying for large-scale industrial use in, for example, the production of calcium oxide, CaO, by roasting in high-temperature rotary furnaces.

(a) Write an equation for the decomposition of calcium carbonate.

(b) Calculate (i) the mass of calcium oxide and (ii) the volume of carbon dioxide, at 25 °C and 1 atm, that could be produced from 1 000 kg of calcium carbonate.

(c) Comment upon *two* environmental concerns regarding the industrial use of limestone.

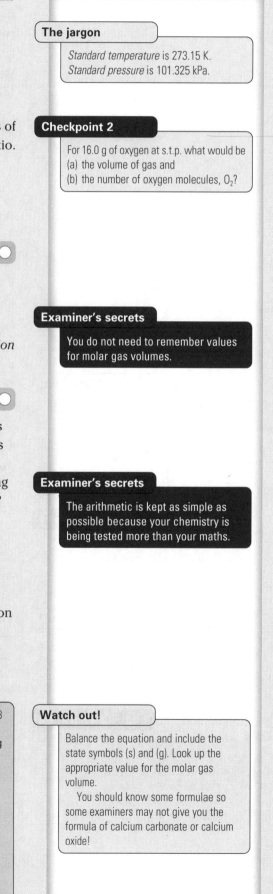

The jargon

Standard temperature is 273.15 K.
Standard pressure is 101.325 kPa.

Checkpoint 2

For 16.0 g of oxygen at s.t.p. what would be
(a) the volume of gas and
(b) the number of oxygen molecules, O_2?

Examiner's secrets

You do not need to remember values for molar gas volumes.

Examiner's secrets

The arithmetic is kept as simple as possible because your chemistry is being tested more than your maths.

Watch out!

Balance the equation and include the state symbols (s) and (g). Look up the appropriate value for the molar gas volume.
  You should know some formulae so some examiners may not give you the formula of calcium carbonate or calcium oxide!

Atomic structure

This section is all about the way electrons are arranged in shells, subshells and orbitals of atoms and ions. You need to understand how to work out these electron arrangements from the patterns in the ionization energies of atoms; but first, what is an ion and what is ionization energy?

Formation of ions

$$Na \rightarrow Na^+ + e^- \text{ and } Mg \rightarrow Mg^{2+} + 2e^-, \text{ etc.}$$
$$Cl + e^- \rightarrow Cl^- \text{ and } O + 2e^- \rightarrow O^{2-}.$$

→ Metal atoms lose electrons to form cations.
→ Non-metal atoms gain electrons to form anions.

Ionization energies

If we supply enough energy (electrical or thermal) to any gaseous atoms they can lose electrons and become ionized.

→ The **molar first ionization**, E_{m1}, of an element is the energy required to remove one mole of electrons from one mole of its gaseous atoms: $X(g) \rightarrow X^+(g) + e^-$.

→ The **molar second ionization**, E_{m2}, of an element is the energy required to remove one mole of electrons from one mole of its gaseous unipositive ions: $X^+(g) \rightarrow X^{2+}(g) + e^-$.

Make sure you can relate the pattern of *successive* ionization energies for an element to the principal electron energy levels in its atoms.

Watch out!

We plot the *logarithm*, E_{mj}, of the ionization energies, not the E_{mj} values.

Checkpoint 1

What is the atomic number and name of the element this pattern refers to?

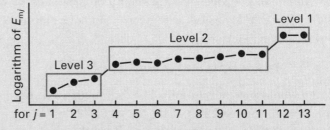

Watch out!

Here we plot the value of the first ionization, E_{m1}, for each of the elements.

You should also be able to relate the pattern of molar first ionization energies for the first 20 elements to the arrangement of electron energies into subsets.

Checkpoint 2

(a) What elements are numbers 2, 10 and 18 at the peaks?
(b) On the chart mark in blue the E_{m1} values for the alkali metals.
(c) On the chart mark in red the E_{m1} values for the halogens.

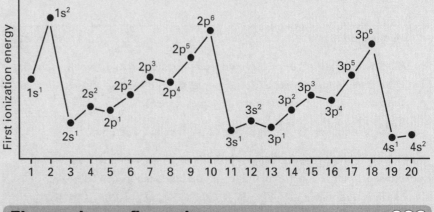

Electronic configurations

The energy levels of electrons in atoms are arranged into sets which are numbered 1, 2, 3, etc. and called shells. This arrangement is called the

electronic configuration which, for potassium and chlorine for example, we write K 2.8.8.1 and Cl 2.8.8.7. The last number in these sequences (1 and 7) refers to the so-called '**valence electrons** in the **outer shell**'. The numbers 2.8.8 refer to the so-called '**inner shell electrons**'.

→ Valence electrons have the lowest ionization energies and are involved in bonding.
→ Inner shell electrons have higher ionization energies and partially shield the outer (valence) shell electrons from the nuclear charge.
→ Atoms of elements in the same group of the periodic table have the same number of valence electrons in their outer shell.
→ The maximum numbers of electrons at each of the first five levels of energy are 2.8.18.32.32.

The jargon

The numbers 1, 2, 3, 4, 5, 6, 7 are called *principal quantum numbers*. The electronic configurations given in data books are for the atoms in their ground states.

Subshells and orbitals ○○○

The principal energy levels of electrons in atoms can be arranged into subsets which are labelled s, p, d, f and called **subshells**.

→ Electrons in the same subshell do not shield each other very much from the attraction of the nucleus.

The jargon

The letters s, p, d and f stem from the words *sharp, principal, diffuse* and *fine* used to describe the lines in spectra.

The subshell energy levels can be arranged into subsets called orbitals. The s-subshell has one orbital, the p-subshell has three, the d-subshell has five and the f-subshell has seven orbitals.

→ No more than two electrons may be in the same orbital.

The jargon

We say that two electrons in the same orbital have *opposite spins*.

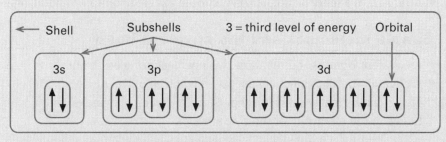

Checkpoint 3

What is the electronic configuration of
(a) Li and Na
(b) F and Br
(c) He, Ne and Ar?

Building up electronic configurations ●●●

Use this plan to work out ground state electronic configurations:

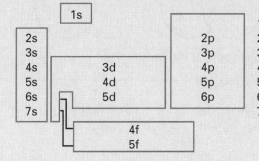

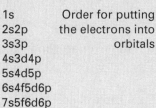

				1s	Order for putting
			2p	2s2p	the electrons into
2s			3p	3s3p	orbitals
3s			4p	4s3d4p	
4s	3d		5p	5s4d5p	
5s	4d		6p	6s4f5d6p	
6s	5d			7s5f6d6p	
7s					
	4f				
	5f				

Watch out!

When you have worked out the number of electrons in each subshell you must write adjacent to each other the subshells with the same principal quantum number. For example:

$1s^22s^22p^63s^23p^64s^23d^3$ is wrong
$1s^22s^22p^63s^23p^63d^34s^2$ is right

Exam question (3 min)   answer: page 23

Write the ground state electronic configurations of atoms with atomic numbers 17, 24 and 33.

Structure and bonding in elements

So what holds things together? Here you will see why nitrogen is a gas with hardly any attraction between one molecule and another, but carbon, in the form of diamond, its next-door-neighbour in the periodic table, is a solid and one of the hardest substances known. You will also see why metals conduct and why non-metals don't. Watch out for the exception to this rule!

Noble gases and van der Waals forces

The jargon

van der Waals forces are weak instantaneous dipole–dipole attractions operating between atoms in all substances.

The noble gas molecules are **monatomic**. Their electron shells are full so the atoms do not combine with each other. However, at very low temperature and high pressure the noble gases condense to liquids in which the atoms are held together by very weak van der Waals forces.

→ The more electrons there are in an atom the more van der Waals forces there will be.

Checkpoint 1

State and explain the trend in b.p. of the noble gases with increasing atomic number from helium to xenon.

Chemists have not yet made any compounds of helium, neon or argon. So the electronic configurations of their atoms and those of the other noble gases are considered to be particularly stable arrangements of electrons. Many (but not all) chemically reactive elements combine so that the electrons in their atoms rearrange to match the electronic configurations of noble gases.

→ Atoms have a tendency to achieve a noble gas electronic configuration.

Simple molecules and the covalent bond

→ A covalent bond is the result of two nuclei *sharing* electrons.
→ A covalent bond forms because the attraction of the nuclei for the electrons is greater than the repulsions between the nuclei and between the electrons of the two atoms.

Watch out!

We represent electrons by dots and crosses only as a convenience to show their origin and help us keep count of the number of valence electrons in each atom. And we usually omit the inner shell electrons.

Hydrogen, the halogens, oxygen and nitrogen molecules are diatomic. We can represent these molecules by 'dot and cross' Venn diagrams:

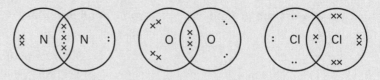

where the overlap shows the sharing of electrons.

:N≡N:	:O=O:	:Cl–Cl:
Triple bond	Double bond	Single bond

The jargon

The — represents a covalent bond of two shared electrons. The : represent a lone pair of non-bonded electrons.

Sulphur and white phosphorus form P_4 and S_8 molecules.

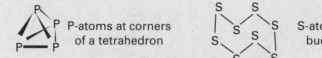

P-atoms at corners of a tetrahedron

S-atoms form a buckled ring

Checkpoint 2

Draw a dot and cross Venn diagram for a molecule of
(a) neon
(b) hydrogen

Simple molecular structures are usually poor conductors that melt and boil at low temperatures, consuming little heat in the process.

Giant covalent structures

Diamond (an allotrope of carbon) and silicon form similar covalently bonded structures that we sometimes call **giant molecules**. The result is an extremely strongly bonded interconnected network of six-membered rings.

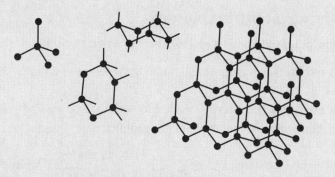

Giant molecular structures are usually poor electrical and thermal conductors that melt at very high temperatures.

Graphite (another allotrope of carbon) forms a giant structure consisting of atoms covalently bonded closely together into large flat sheets of flat interconnected hexagons.

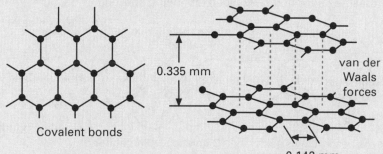

Covalent bonds

0.335 mm

van der Waals forces

0.142 mm

Within each sheet some of the valence electrons (not used in the covalent bonding) are free to move. These mobile electrons give graphite its good electrical conductivity. The ability of the sheets to slide over one another gives graphite its slippery lubricating property.

Metallic structures

Metals are elements or mixtures (alloys) of elements.

→ A metal consists of a lattice of mutually repelling positive ions held together by their attraction for a 'sea' of mobile electrons.
→ Most metals are **close-packed** structures (hcp or ccp – coordination 12) but alkali metals are soft **body-centred** structures.

The jargon

Allotropes are different forms of the same element having different physical (and sometimes chemical) properties.

Watch out!

An extremely pure graphite crystal is hard and its electrical conductivity is low in one direction.

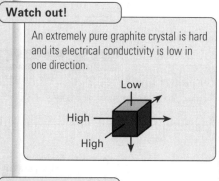

Low

High

High

The jargon

The mobile electrons in graphite and in metals are called *delocalized electrons*.

The jargon

An *alloy* is a solid solution of two or more metals. Sometimes a non-metal forms part of an alloy.

Exam question (8 min)   answer: page 23

(a) State three characteristic properties of metals.

(b) How does the addition of carbon alter the physical properties of pure iron?

(c) Sketch a unit cell of the alkali metal structure, state the coordination number and name the type of structure.

Structure and bonding in covalent compounds

The bonding in binary non-metal compounds is covalent but bond polarization may produce some ionic character. You can use the valence shell electron pair repulsion (VSEPR) theory to predict the shapes of simple covalent molecules.

Simple covalent molecules ●●●

A covalent bond exists between two atoms when they share electrons.

→ **Bond length** is the average distance between the nuclei of the two atoms.
→ **Bond energy** is the energy needed to separate completely the atoms in the molecules of one mole of compound.

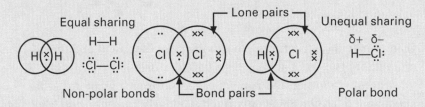

→ A dative (or coordinate) bond is a covalent bond in which one of the atoms has supplied *both* electrons being shared.

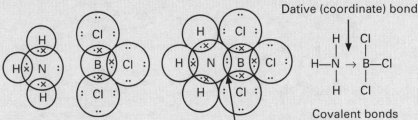

Lewis base   Lewis acid   Dative (coordinate) bond

Polar bonds ●●●

In a diatomic hydrogen (or chlorine) molecule the two identical nuclei must attract the bonded electron pair of electrons equally strongly. In a molecule of HCl the chlorine nucleus (albeit shielded by two inner shells of electrons) attracts the electrons more strongly than does the hydrogen nucleus. This makes the H-atom *slightly* positive and the Cl–atom *slightly* negative. Consequently, the H–Cl bond is **polar** and the HCl molecule has a **dipole moment**.

Electronegativity

The **Pauling electronegativity index** (Np) is a measure of how strongly an atom in a compound attracts electrons in a bond.

Trends in electronegativity index values in the periodic table:

Increasing electronegativity →

C (2.5)	N (3.0)	O (3.5)	F (4.0)
Si (1.8)	P (2.1)	S (2.5)	Cl (3.0)
	As (2.0)	Se (2.4)	Br (2.8)
		Te (2.1)	I (2.5)

The greater the difference in the electronegativities of two atoms, the more polar is the covalent bond between the two atoms.

Don't forget!

For any two given atoms the bond gets shorter and stronger as it changes from $-$ to $=$ to $\equiv$.

Watch out!

Dots and crosses help to show which valence electrons come from which atom *but* once a bond is formed the electrons are indistinguishable.

Checkpoint 1

Draw a dot and cross diagram for a molecule of (a) methane, (b) carbon dioxide and (c) iodine chloride.

The jargon

A *Lewis acid* is a molecule or ion that can accept a pair of electrons to form a covalent bond. What is a Lewis base? See page 63 for the meaning of a Brønsted–Lowry acid and base.

Examiner's secrets

You don't need to remember Np values but you do need to learn these trends. They are really important.

Checkpoint 2

Which halogen is most electronegative and which hydrogen halide is the least polar?

Polar molecules

Oxygen is more electronegative than carbon and more electronegative than hydrogen. Consequently the bonds between oxygen and these elements are polarized so the O-atom is slightly negative.

Why is water a polar molecule while CO_2 is not?

In carbon dioxide the two CO bonds are equally polar but their dipoles act in opposite directions and cancel each other out.

In water the dipoles of the two HO bonds act in similar directions and reinforce each other.

Don't forget!

Polar molecules means polar bonds but polar bonds don't always mean polar molecules. The shape of the molecule is very important.

Shapes of molecules ●●●

You should be able to predict the approximate angles between bonds and the shape of simple molecules using rules based on the **VSEPR** theory.

1 Write formulae to show all electron (− bonded and : lone) pairs.

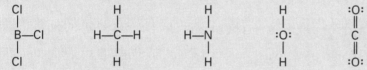

2 Assume the electron pairs (bonded and lone) move equally as far apart as possible from each other but treat double bonds as single pairs.

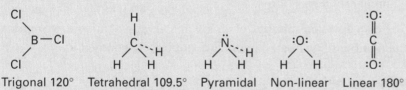

Trigonal 120°   Tetrahedral 109.5°   Pyramidal   Non-linear   Linear 180°

3 Adjust any bond angles affected by the following rule for repulsion between bonded (bp) and non-bonded (nbp) electron pairs:

nbp.nbp repulsion > nbp.bp repulsion > bp.bp repulsion

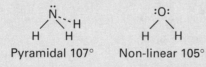

Pyramidal 107°   Non-linear 105°

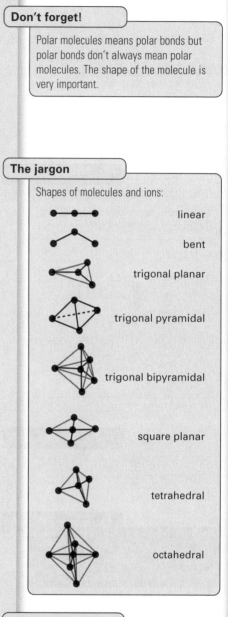

The jargon

Shapes of molecules and ions:

linear

bent

trigonal planar

trigonal pyramidal

trigonal bipyramidal

square planar

tetrahedral

octahedral

Watch out!

Elements like silicon, phosphorus and sulphur can extend their outer valence beyond 8.

Exam question (10 min)   answer: page 24

(a) Draw dot and cross diagrams to represent the following covalent molecules: (i) H_2S, (ii) NI_3, (iii) PCl_5, (iv) Al_2Cl_6.

(b) For each of the molecules in (a) predict and draw the shape of the molecule and indicate its likely polarity, if any.

Examiner's secrets

There will be a separate mark if the dative bonds in the Al_2Cl_6 diagram are represented correctly.

15

Structure and bonding in ionic compounds

The bonding in binary metal-non-metal compounds is ionic but cations may polarize anions to produce some covalent character. You can use the octet rule and electronic configurations to predict the formula of simple compounds.

Simple ionic bonding

Ionic bonding is the result of electrons being transferred from one atom to another and the ions packing together into a crystal **lattice**.

Predicting simple ionic formulae

Example: The formula of rubidium oxide

The atomic number of rubidium is 37 and of oxygen is 8. Their ground state electronic configurations are

$$\text{Rb } 2.8.18.8.1 \text{ and O } 2.6$$

Predict Rb to lose one electron and O to gain two electrons:

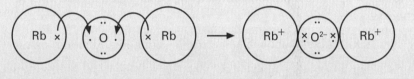

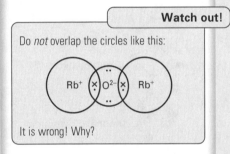
Ionic crystal structures

You should be able to

→ draw diagrams and describe the unit cells of CsCl and NaCl
→ work out the number of ions in each unit cell
→ deduce the empirical formula from the ratio of the coordination numbers of each ion

The **coordination number** of an ion in a crystal of an ionic compound is the number of nearest, equidistant, oppositely charged ions.

Caesium chloride
Double simple cubic

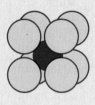

= Cl⁻
= Cs⁺

Coordination no. $Cs^+ = 8$
Coordination no. $Cl^- = 8$
Ratio of $Cs^+ : Cl^- = 1 : 1$
Hence formula is CsCl

Sodium chloride
Face-centred cubic

What is the coordination no. $Na^+ =$ ___
What is the coordination no. $Cl^- =$ ___
What is the ratio of $Na^+ : Cl^- =$ ___ : ___
Hence formula is NaCl

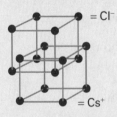

= Cl⁻
= Na⁺

Crystal systems

There are seven different crystal systems: cubic, hexagonal, monoclinic, orthorhombic, rhombohedral, tetragonal and triclinic. The type of

crystal structure adopted by an ionic compound is governed by the relative numbers, shapes and sizes of the cations and anions.

→ **Radius ratio** is the radius of the smaller ion divided by the radius of the larger ion.

Hydrated crystal structures ●●○

Cations attract the negative (oxygen) end of water molecules and anions attract the positive (hydrogen) end of water molecules.

→ Many ionic compounds form hydrated crystals in which the water molecules are held in the lattice by **ion–dipole forces**.
→ Hydrated ionic compounds melt at much lower temperatures than the corresponding anhydrous compounds.
→ Less heat is required to melt hydrated ionic compounds which often decompose by loss of water.

Complex cations ●●○

Transitional metal cations form complex ions in which non-metal ligands are attached to the metal ion by **dative bonds**.

Polyatomic non-metal ions ●●○

Many molecules may form anions (or cations) by acting as acids (or bases) and losing (or gaining) protons.

You should be able to draw dot and cross diagrams and predict the shape of some simple anions and cations.

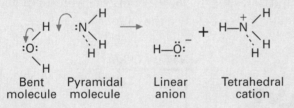

| Bent molecule | Pyramidal molecule | Linear anion | Tetrahedral cation |

Delocalized bonding

When you use the **octet** and **VSEPR** theories to predict the shapes of certain molecules or ions and you find that you can write more than one possible arrangement of the bonds, the actual structures are probably delocalized.

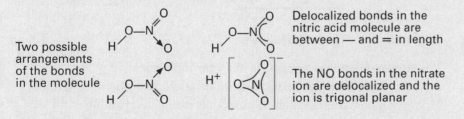

Two possible arrangements of the bonds in the molecule

Delocalized bonds in the nitric acid molecule are between — and = in length

The NO bonds in the nitrate ion are delocalized and the ion is trigonal planar

---

Checkpoint 4

(a) Use the following ionic radii (in nm) to calculate the radius ratios in caesium chloride and sodium chloride:
Cs$^+$ = 0.170; Na$^+$ = 0.102; Cl$^-$ = 0.180
(b) What crystal system is adopted by (i) caesium chloride, (ii) sodium chloride?

The jargon

Ion–dipole interaction is the attraction between an ion and a polar molecule. *Anhydrous* means no water of crystallization.

The jargon

A *ligand* is a molecule or anion with one or more atoms having a lone pair of electrons for donation to form a dative bond.

Links

See page 121: ligands.

The jargon

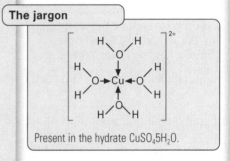

Present in the hydrate CuSO$_4$5H$_2$O.

Exam question (10 min)   answer: page 24

(a) Draw a dot and cross diagram for (i) sodium hydride, (ii) ammonium chloride, (iii) the oxonium ion.
(b) Suggest, in terms of bonding, why the melting point of aluminium fluoride is much higher than the melting point of aluminium chloride.
(c) Deduce the structure and shape of the carbonate ion.

Intermolecular forces

We know that a molecule is held together by strong covalent bonds but what holds one molecule to another? Why are they attracted at all? Why is water a liquid? Why does the boiling point of the halogens increase down the group? Read on!

van der Waals forces

van der Waals forces are weak *short-range* forces between atoms and molecules arising from the attraction between dipoles. These forces arise from the movement of the electrons in relation to the nuclei which produces *weak instantaneous dipoles* that attract one another.

→ The more electrons there are in an atom, the more instantaneous dipole–dipole attractions there will be.
→ van der Waals forces account for the increase in boiling point with increasing molar mass of **noble gases** and **halogens**.

Dipole–dipole attractions

Dipole–dipole forces are the attractions between the positive end of one *permanently* polar molecule and the negative end of another *permanently* polar molecule.

Ion–dipole interactions

Many ionic compounds dissolve well in water and other polar solvents because the energy released by the formation of the ion–dipole interactions compensates for the **lattice energy** consumed in overcoming the electrostatic forces between the cations and anions.

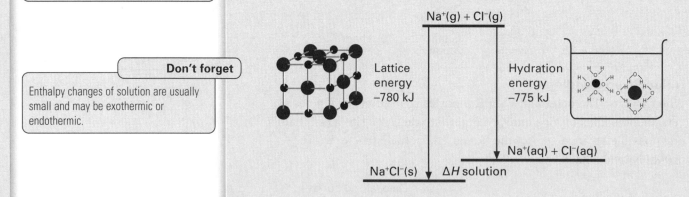

We picture the process as follows:

1 Cations and anions attract polar water molecules.
2 Water molecules get between the ions and lessen their attractions.
3 Water molecules completely surround cations and anions.
4 Hydrated ions leave the crystal and mix with the water.

Checkpoint 1

Suggest why the b.p. of iodine chloride is higher than the b.p. of bromine.

The jargon

Hydration energy is the enthalpy change for the formation of one mole of the aqueous ions from ions in the gaseous state.

Don't forget

Enthalpy changes of solution are usually small and may be exothermic or endothermic.

Checkpoint 2

(a) What is the molar enthalpy change of solution of sodium chloride?
(b) Does salt dissolve in water exothermically or endothermically?

Hydrogen bonding

●●●

A hydrogen bond (···) is a weak bond between a very electronegative atom (X = N, O or F) and a hydrogen atom bonded to a very electro-negative atom (Y = N, O or F). Thus −X···H−Y.

→ Hydrogen bonding is stronger than van der Waals forces and permanent dipole–dipole attractions but weaker than covalent bonding.

You should be able to explain the exceptionally high boiling points of ammonia, water and hydrogen fluoride (compared to the values for the other hydrides in each group) by hydrogen bonding.

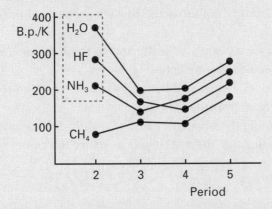

Hydrogen bonding is a major factor in determining the structure and properties of water, hydrated salts, carbohydrates and proteins.

Dimerization of acids

Hydrofluoric acid is atypical in being a weak acid ($pK_a = 3.3$) because it **dimerizes** in water, due to hydrogen bonding. This decreases the number of protons donated to the water molecules:

$$H_2O(l) + H−F···H−F(aq) \rightarrow H_3O^+(aq) + [F···H−F]^-(aq)$$

In pure crystals or anhydrous liquids, carboxylic acids form **dimers**, due to hydrogen bonding.

Example:

$$H−C\overset{O···H−O}{\underset{O−H···O}{}}C−H$$

Methanoic acid
m.p./°C = 8.3
b.p./°C = 100.5
M_r (dimer) = 92

answer: pages 24–5

Exam question (10 min)

(a) Explain each of the following.

(i) The boiling point of hydrogen fluoride, HF, is the highest of all the hydrogen halides.

(ii) In non-aqueous media, ethanoic acid exists as dimeric molecules $(CH_3CO_2H)_2$.

(iii) Water in the solid state is less dense than water in the liquid state.

(b) (i) State the type of ionic crystal structure and the coordination number of the cation in caesium chloride.

(ii) Explain why the answer in (b) (i) would be different for sodium chloride.

The jargon

A *hydrogen bond* may be explained by saying the highly electronegative atom Y polarizes the H−Y bond so much that the nucleus of the H-atom is exposed to attraction by a lone pair on atom X.

Checkpoint 3

(a) What are the formulae and names of the twelve other hydrides whose b.p.s are shown on the chart?

(b) Suggest why water has a higher b.p. than hydrogen fluoride. (Hint: for each hydride, find the average number of hydrogen bonds between a pair of its molecules.)

Watch out!

Dimerization can produce unexpected properties. Note the similarity of benzene (molecule and its properties) to methanoic acid.

Benzene
m.p./°C = 5.4
b.p./°C = 80.0
$M_r = 78$

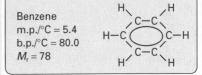

Examiner's secrets

Marks are usually given for diagrams and for clear labels. A clearly labelled diagram is all you need to get full marks for (a) (ii) for example.

Watch out!

Take care not to spend too much time drawing and colouring pretty diagrams!

Gases

This section is about gases and the laws that describe their behaviour. You need to understand the ideal gas equation and know how to use it.

Kinetic molecular theory of gases ●●●

The jargon

Kinetic comes from the Greek *kinetikos*, to move.

We explain the physical properties of gases by picturing tiny molecules in constant rapid and random motion colliding with themselves and with the walls of their container. In this theory for an **ideal gas** we make the following important assumptions.

→ The volume of the molecules themselves is negligible compared with the volume of the containing vessel they occupy.
→ The molecules are not attracted to each other or to the walls of the container.
→ All collisions are perfectly elastic and so the total kinetic energy of all the molecules is constant.
→ The mean kinetic energy of the molecules is directly proportional to the Kelvin temperature of the gas.
→ The collision of the molecules with the walls of their container is responsible for the observable gas pressure (force per unit area).

The gas laws ●●●

Checkpoint

(a) Sketch the graph of volume against pressure for a given mass of ideal gas at constant temperature.
(b) Sketch the graph of volume against Kelvin temperature for a given mass of ideal gas at constant pressure.

→ Boyle's law: $pV =$ a constant

For a given mass of gas, at constant temperature, the volume (V) is inversely proportional to the pressure (p).

→ Charles' law: $V/T =$ a constant

For a given mass of gas at constant pressure, the volume (V) is directly proportional to the Kelvin temperature (T).

→ Avogadro's law: $V \propto n$

For a gas at constant temperature and pressure, the volume (V) is proportional to the number of molecules (n).

The general gas equation ●●●

The jargon

The value of the gas constant R is usually given as 8.314 J K^{-1} mol^{-1}. To be consistent with the units we need to use volume (V) in m^3, pressure (p) in N m^{-2} (= Pa) and temperature (T) in K. The units of molar mass (M) are g mol^{-1}.

The three laws above can be combined into one equation.

→ ideal gas equation: $pV = nRT$
→ n (number of moles) $= m/M$

m is the mass of gas in grams and M is its molar mass.

Using the ideal gas equation

If we can measure the volume of a known mass of a compound at a known temperature and pressure, then we can determine the molar mass of the compound.

Don't forget!

When you make measurements on a gas or vapour and do your calculations using $pV = nRT$ remember that you are assuming ideal behaviour.

If the gas dissociates, then the value of M_r (the relative molecular mass) can tell us the degree of dissociation α at the given temperature and pressure. This in turn can lead to a value for K_p, the equilibrium constant for the dissociation.

Example: Calculate the molar mass of a volatile organic liquid X given that the volume of 0.567 1 g of its vapour is 145.2 cm³ at 100 °C and 101.3 kPa.

The general gas equation $pV = (m/M)RT$ can be rewritten in the form

$$M = mRT/pV$$

Substituting for $m = 0.567\ 1$ g, $V = 1.452 \times 10^{-4}$ m³, $p = 101\ 300$ Pa, $T = 373$ K gives

$$M = (0.567\ 1 \times 8.134 \times 373)/(101\ 300 \times 1.452 \times 10^{-4})$$
$$= 119.5 \text{ g mol}^{-1}$$

Dalton's law of partial pressures ●●●

Make sure you understand this law. You often need to use partial pressure when calculating equilibrium constants.

→ In a mixture of two or more gases, the partial pressure of each gaseous component is the pressure which that gas would exert if it alone occupied the volume taken up by the mixture of gases.
→ The total pressure (P) of a gaseous mixture equals the sum of the individual partial pressures (P) of the components: $P = \Sigma p$.
→ The partial pressure of any component in a mixture of gases is given by $p = (n_c/n_t) \times P$.

Real gases ●●●

At very high pressure and low temperature the volume of the molecules themselves becomes significant. Furthermore, the experimental fact that all gases can be liquefied under these conditions shows that there must be forces of attraction between molecules.

→ All gases deviate considerably from the ideal behaviour at high pressures and low temperatures.

van der Waals' equation
J. C. van der Waals, a Dutch chemist, proposed the following gas equation to allow for non-ideal behaviour:

$$(p + a/V_m^2)(V_m - b) = RT$$

The constant a allows for the intermolecular forces of attraction and the constant b allows for the volume of the molecules themselves.

Exam question (7 min)   answer: page 25

(a) Calculate the volume occupied by 0.500 g of propanone at 100 °C and 101.3 kPa pressure.

(b) The equation for the dissociation of dinitrogen tetraoxide is

$$N_2O_4(g) \rightleftharpoons 2NO_2(g)$$

The dinitrogen tetraoxide is 20% dissociated at a temperature of 28 °C and 101.3 kPa pressure. Calculate the volume occupied by one mole of dinitrogen tetraoxide under these conditions.

[The molar gas volume is 22.4 dm³ at s.t.p.]

Examiner's secrets

You will always be given the gas constant value, usually as $R = 8.314$ J K⁻¹mol⁻¹. Notice the units in J (= joules), *not* KJ (= kilojoules). This should remind you to be sure to use the correct units for p, V and T.

Links

See pages 53–71: equilibria.

The jargon

n_c is the number of moles of the component gas. n_t is the total number of moles of gas in the mixture. n_c/n_t is called the mole fraction.

Examiner's secrets

Make sure you learn your organic formulae. You should at least know the name and formula of the first member of the important homologous series. Propanone is the simplest ketone. Its traditional name is acetone.

Structured exam question

answer: page 25

(a) Write the ground state electronic configurations, in terms of s and p electrons, for each of the following isolated species:

(i) a chlorine atom ....

(ii) a calcium ion ....

(iii) a sulphide ion ....

(b) Draw dot and cross diagrams to represent the bonding in each of the following structural units:

(i) ammonia, NH_3

(ii) carbon dioxide, CO_2

(iii) sodium oxide, Na_2O

(c) (i) Describe the difference between the structures of sodium chloride and caesium chloride. Your answer may be either a written description or clearly labelled diagrams.

(ii) Give a reason why sodium chloride and caesium chloride have different structures.

....

....

(10 min)

Examiner's secrets

If you are drawing a diagram for *simple* ions do *not* show dots and crosses being shared. You won't get a mark.

Examiner's secrets

Simple clearly labelled neat drawings are better than long illegible muddled written descriptions.

Examiner's secrets

They have because they have is true but not the answer. Think ion sizes.

Answers
Atoms, Ions and molecules

Formulae and equations

Checkpoints

1 SO_2
2 (a) K^+I^- (b) $Ca^{2+} SO_4^{2-}$
   (c) $(Li^+)_2 O^{2-}$ (d) $Ba^{2+}(OH^-)_2$
   (e) $(Al^{3+})_2(SO_4^{2-})_3$
3 (a) Ar   (b) Cl_2 (c) O_3
4 (a) ICl   (b) SCl_2 (c) CO   (d) NI_3

Exam question

(a) $KOH + HNO_3 \rightarrow KNO_3 + H_2O$
    $H^+(aq) + OH^-(aq) \rightarrow H_2O(l)$
(b) $Ba(OH)_2 + 2HCl \rightarrow BaCl_2 + 2H_2O$
    $H^+(aq) + OH^-(aq) \rightarrow H_2O(l)$

Atoms and isotopes

Checkpoints

1 Carbon C
2 $^{13}_{6}C$
3 44 and 46
4 (a) 2p, 2n
   (b) (i) Z decreases by 2 and A decreases by 4.
       (ii) Z increases by 1 and A remains the same.
5 Carbon contains small amounts of C-13 and C-14 isotopes.

Exam question

(a) (i) Cl-35 17p   Cl-37 17p   (ii) Cl-35 18n   Cl-37 20n
(b)

35	35.5	36	36.5	37
	1		3	

three Cl-35 : one Cl-37 = 75% Cl-35 and 25% Cl-37
Or A_r(Cl) = 35.5; if x = percentage of Cl-35, then
$35.5 = [35x + 37(100 - x)]/100$; so $x = 75\%$.

Amount of substance and the mole

Checkpoints

1 (a) (i) 1.01 g   (ii) 14.0 g   (iii) 48.0 g
   (b) 63.01 g mol^{-1}
2 (a) 11.2 dm^3
   (b) 3×10^{23}

Exam question

(a) $CaCO_3 \rightarrow CaO + CO_2$
(b) (i) 560 kg of calcium oxide
    (ii) 2.4×10^5 dm^3
(c) E.g. Scarring of landscape by mining; production of dust by processing; release of CO_2 on roasting to CaO.

Atomic structure

Checkpoints

1 13, aluminium
2 (a) He, Ne and Ar
   (b) s^1 elements marked blue
   (c) p^5 elements marked red
3 (a) Li $1s^2 2s^1$ and Na $1s^2 2s^2 2p^6 3s^1$
   (b) F $1s^2 2s^2 2p^5$ and Br $1s^2 2s^2 2p^6 3s^2 3p^6 3d^{10} 4s^2 4p^5$
   (c) He $1s^2$, Ne $1s^2 2s^2 2p^6$ and Ar $1s^2 2s^2 2p^6 3s^2 3p^6$

Exam question

$1s^2 2s^2 2p^6 3s^2 3p^5$
$1s^2 2s^2 2p^6 3s^2 3p^6 3d^5 4s^1$
$1s^2 2s^2 2p^6 3s^2 3d^{10} 4s^2 4p^3$

Structure and bonding in elements

Checkpoints

1 As the atomic number increases, the number of electrons per atom increases and so the van der Waals forces increase. Consequently, more energy is needed to separate the liquid molecules and turn them into gas.
2 (a)

   (b)

Exam question

(a) Malleable; lustrous; good electrical conductivity
(b) Iron becomes harder and less ductile
(c) 8; body-centred cubic

Structure and bonding in covalent compounds

Checkpoints

1 (i)

(ii)

(iii)

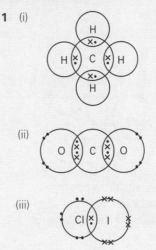

2 Fluorine and hydrogen iodide

Exam question

(a) (i)

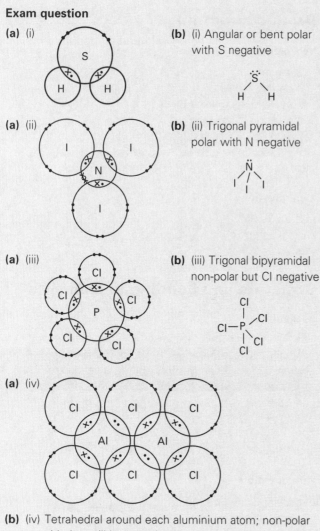

(b) (i) Angular or bent polar with S negative

(a) (ii)

(b) (ii) Trigonal pyramidal polar with N negative

(a) (iii)

(b) (iii) Trigonal bipyramidal non-polar but Cl negative

(a) (iv)

(b) (iv) Tetrahedral around each aluminium atom; non-polar chlorine will be negative

Structure and bonding in ionic compounds

Checkpoints

1 Cs⁺ and F⁻

2
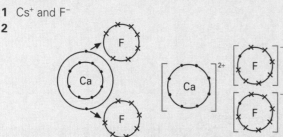

3 (a) Each ion is surrounded by six or eight oppositely charged ions which are held together in the lattice by strong electrostatic forces and much energy is needed to overcome them.

(b) In the solid the ions are in fixed positions but in the liquid the ions are free to move and conduct electricity.

---

4 (a) CsCl 0.94   NaCl 0.57

(b) (i) CsCl cubic (double simple interlocking)
   (ii) NaCl cubic (face-centred interlocking)

Exam question

(a) (i)

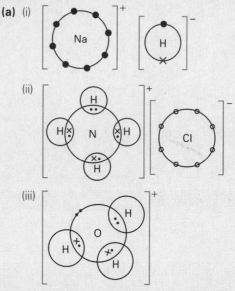

(ii)

(iii)

(b) Aluminium fluoride will have considerable ionic character whereas aluminium chloride is covalent.

(c) The bonding in the carbonate ion will be delocalized (the 2– charge shared equally by the three O-atoms), the three carbon–oxygen bonds will be identical and the C-atom will *not* have a lone pair. Hence the ion will be trigonal planar.

Intermolecular forces

Checkpoints

1 The difference in electronegativity produces a permanent dipole in I–Cl. The Br–Br cannot have a permanent dipole.

2 (a) This is the enthalpy change when one mole of sodium chloride is dissolved in water to a given dilution. In data books the values are usually given to infinite dilution.

(b) Endothermically.

3 (a) Hydrogen sulphide H_2S
   Hydrogen selenide H_2Se
   Hydrogen telluride H_2Te
   Hydrogen chloride HCl
   Hydrogen bromide HBr
   Hydrogen iodide HI
   Phosphine PH_3
   Arsine AsH_3
   Stibine SbH_3
   Silane SiH_4
   Germane GeH_4
   Stannane SnH_4

(b) The average number of hydrogen bonds between a pair of water molecules is two but it is only one between a pair of HF or NH_3 molecule.

Exam question

(a) (i) Fluorine is the most electronegative element so the hydrogen bonding between HF molecules outweighs

the increase in van der Waals forces with increasing molar mass from HCl to HI. Extra energy is needed to overcome the hydrogen bonding before the $H_2F_2(l)$ molecules can separate and become gaseous.

(ii) The dimer is held together by hydrogen bonds:

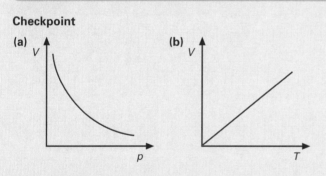

(iii) In ice there is an open lattice structure which is held together by hydrogen bonding. On melting, the hydrogen bonding begins to break and the open structure starts collapsing. The water molecules begin to pack more closely together in the liquid state so the density increases until it reaches a maximum at 4 °C.

(b) (i) Two interlocking simple cubic systems giving a coordination number of 8 : 8.

(ii) The radius ratio for sodium chloride is smaller than that for caesium chloride, so the sodium and chloride ions can form a face-centred cubic lattice with a coordination number of 6 : 6.

Gases

Checkpoint

(a)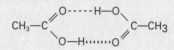

(b)

Exam question

(a) Molar mass of propanone is 58.08 g mol^{-1} so 0.500 g is 0.500/58.08 = 0.008 61 mol propanone. At 273 K the volume of one mole is 22 400 cm^3 so for 0.008 61 mol V is 22 400 × 0.008 61 = 192.9 cm^3 At 373 K the volume is 192.9 × (373/273) = 263.4 cm^3 so volume of propanone will be 263 cm^3.

(b) One mole of ideal gas at 28 °C and 101.3 kPa would be 22.4 × (301/273) = 24.7 dm^3.
When 20% of the $N_2O_4(g)$ dissociates the total amount of gas molecules increases by 20% (because for every $N_2O_4(g)$ molecule lost, two $NO_2(g)$ molecules are gained). So volume increases by 20% from 24.7 dm^3 to 29.6 dm^3.

Structured exam question

(a) (i) a chlorine atom $1s^2 2s^2 2p^6 3s^2 3p^5$
(ii) a calcium ion $1s^2 2s^2 2p^6 3s^2 3p^6 4s^2$
(iii) a sulphide ion $1s^2 2s^2 2p^6 3s^2 3p^6$

(b) (i)   ammonia, NH_3

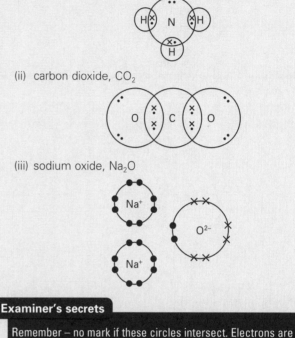

(ii) carbon dioxide, CO_2

(iii) sodium oxide, Na_2O

(c) (i) In sodium chloride, each ion is surrounded by six of the oppositely charged ions. Sodium chloride has a double interlocking face-centred cubic structure.

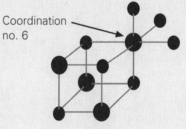

Coordination no. 6

Na^+ ion Cl^- ion

In caesium chloride, each ion is surrounded by eight of the oppositely charged ions. Caesium chloride has a double interlocking simple cubic structure.

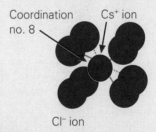

Coordination no. 8 Cs^+ ion

Cl^- ion

(ii) The caesium ion is much larger than the sodium ion and the radius ratio for caesium chloride is larger than for sodium chloride.

Energetics

In this chapter you will meet the idea of an enthalpy change ΔH, a heat change at constant pressure. When you see the symbol $\Delta H^\ominus$ you will know this refers to a change under standard conditions of temperature, pressure, etc. For A-level you should know how to measure heats of reactions such as combustion and neutralization. Make sure you understand and can use Hess's law to calculate enthalpy changes of formation, bond enthalpies and lattice energies. These calculations are very popular in exams.

Exam themes

→ Calculate enthalpy changes from experimental data. Use temperature rises and the equation '$H = mc\Delta T$' to calculate the heat produced in a calorimeter.

→ Use Hess's law in thermochemical calculations

→ Calculate enthalpy changes using ΔH_f (enthalpy of formation), ΔH_c (enthalpy of combustion) and bond energy data.

→ Draw and use Born–Haber cycles

→ Do calculations involving ΔG and ΔS and predict the thermodynamic feasibility of reactions (only some exam boards)

Topic checklist

○ AS ● A2	AQA	CCEA	EDEXCEL	OCR	WJEC
Measuring enthalpy changes	○	○	○	○	○
ΔH and Hess's law	○	○	○	○	○
Bond energies	○	○	○	○	○
Lattice energies	●	●	●	●	●
Free energy and entropy	●				

Measuring enthalpy changes

Chemists use a bomb calorimeter to measure heats of combustion at constant volume. You could measure heats of combustion at constant pressure with a flame calorimeter and heat changes involving solutions with a polystyrene cup. In every experiment the basic principles are the same; what are they?

Basic principles ●●●

To determine a heat change (δh) involving a certain amount of substance, we measure the temperature change (δT) in the contents of a calorimeter with heat capacity C: $\delta h = \delta T \times C$. We can write this equation in words and with units:

$$(\text{heat change/kJ}) = (\text{temperature change/K}) \times (\text{heat capacity/kJ K}^{-1})$$

units: kilojoules   Kelvin   kilojoules per Kelvin

K cancels K^{-1}

We use the measured δh to calculate ΔH, the molar heat change.

→ An enthalpy change ΔH is a heat change **at constant pressure**.
→ For **exothermic** processes ΔH has a *negative* value.
→ For **endothermic** processes ΔH has a *positive* value.

Using a simple calorimeter ●●●

Example: Measuring the **enthalpy change of neutralization**

1 React 50 cm^3 hydrochloric acid with 50 cm^3 aqueous sodium hydroxide, each solution of concentration 1.0 mol dm^{-3}.
2 Measure the rise in temperature of the resulting aqueous sodium chloride.
3 Ignore the heat capacity of the cup and assume the specific heat capacity of the aqueous sodium chloride to be that of water.
4 Multiply 4.2 J K^{-1} g^{-1} by the mass (g) of the sodium chloride solution to find its heat capacity.
5 Calculate the **molar** enthalpy change of neutralization.

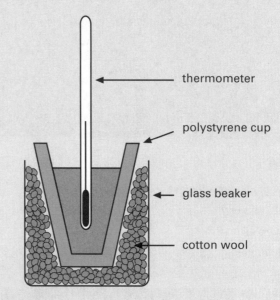

thermometer

polystyrene cup

glass beaker

cotton wool

Using a flame calorimeter

A flame calorimeter measures heats of combustion *at constant pressure*.

Example: Measuring the enthalpy change of combustion of an alcohol.

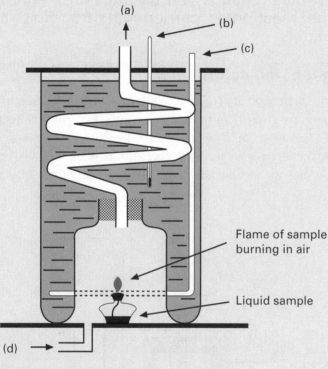

(a)

(b)

(c)

Flame of sample burning in air

Liquid sample

(d)

To determine the heat capacity (C) of the calorimeter you could, for example, measure the loss in mass of a sample of propan-2-ol burnt in the calorimeter and the resulting temperature rise. You could then calculate the value of C using the expression

$$\left\{\frac{\Delta H}{\text{kJ mol}^{-1}} \times \left[\left(\frac{\delta m}{\text{g}}\right) \div \left(\frac{M}{\text{g mol}^{-1}}\right)\right]\right\} \div \left(\frac{\delta T}{\text{K}}\right)$$

In practice it is quite difficult to obtain a steady, smokeless flame simply by adjusting the height of the burner's wick.

Exam question (8 min) answer: page 39

In an experiment to measure the heat of combustion of butan-1-ol ($M_r = 74.1$) a student first determined the heat capacity of a flame calorimeter using propan-2-ol ($M_r = 60.1$) whose enthalpy change of combustion is 2 006 kJ mol^{-1}. In the calibration experiment the combustion of 0.149 g propan-2-ol produced a temperature rise of 5.42 °C. In a second experiment the combustion of 0.152 g butan-1-ol produced a temperature rise of 5.21 °C.

(a) Calculate (i) the heat capacity of the flame calorimeter and (ii) the student's value for the heat of combustion of the butan-1-ol.

(b) The standard enthalpy of combustion of butan-1-ol is 2 676 kJ mol^{-1}.
   Comment on the accuracy of the student's experiment and suggest two ways of improving it.

The jargon

Biologists call this a 'food calorimeter'. We use it to measure heats of combustion at constant pressure.

Watch out!

Don't confuse a flame calorimeter with a bomb calorimeter: see page 30.

Checkpoint 2

Fill in the four missing diagram labels (a) to (d) to explain the purpose of each part being labelled.

The jargon

ΔH is the molar heat of combustion of propan-2-ol, δm is its loss in mass, δT is the rise in temperature and M is its molar mass.

Checkpoint 3

What would be the units of C?

Watch out!

A smoky flame is a sure sign of incomplete combustion but even with a clear blue flame some of the carbon in an organic compound may burn only to carbon monoxide.

Checkpoint 4

Discuss briefly two environmental problems associated with the combustion of organic compounds.

ΔH and Hess's law

Molar heat changes vary with temperature, pressure and the concentration of the reactants. In order to compare different reactions, we measure heat changes under standard conditions. According to Hess's law, the standard heat change will be the same regardless of how the reaction is carried out. But first, another calorimeter.

Using a bomb calorimeter

We find the heat capacity (C) of a bomb calorimeter by measuring the temperature rise (δT) produced by a known amount of heat (δh): $C = \delta h / \delta T$. This heat could be supplied either by a measured amount of electrical energy or by burning a measured amount of a reference substance such as benzoic acid with a known heat of combustion.

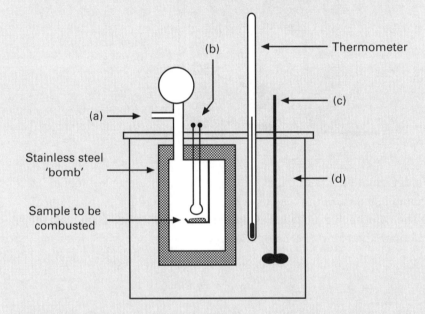

A bomb calorimeter measures what we call **internal energy changes** of combustion (ΔU_c). Internal energy change is used to calculate the enthalpy change ΔH_c.

→ *Combustion* is always *exothermic* so ΔH_c is always *negative*.

Molar enthalpy change

The amount of substance used in calorimeters is quite small: much less than one mole. The heat change found for a measured but small amount of reaction can be used to calculate the molar enthalpy change.

→ The **molar enthalpy change** of combustion is the heat evolved when one mole of substance is completely burnt in oxygen.

Standard molar enthalpy change
Molar enthalpy change is a more precise term than molar heat change but it is still not precise enough. The ΔH value for a reaction may vary with temperature, pressure, concentration and physical state of reactants and products.

The jargon

Bomb refers to the fact that the solid or liquid sample is ignited electrically in an atmosphere of pure oxygen at a pressure of about 20 atm.

Checkpoint 1

Fill in the four missing diagram labels (a) to (d) to explain the purpose of each part being labelled.

The jargon

Internal energy change ΔU is the heat change at constant volume. We calculate ΔH from ΔU using $\Delta H = \Delta U + \Delta n RT$ where R is the gas constant, T is the absolute temperature and Δn is the number of moles of gaseous products minus the number of moles of gaseous reactants.

Checkpoint 2

Calculate the molar enthalpy of combustion of octane, C_8H_{18}, given that 0.114 g of the liquid gives off 5.47 kJ of heat when burnt completely to carbon dioxide and water.

Watch out!

Molar can refer to the 'amount of change' specified by the chemical representing the amounts of substances involved in the change.

→ $\Delta H^{\ominus}_{298}$ represents a **standard molar enthalpy change** for a process in which all substances are in their most stable forms at a temperature of 298 K (25 °C) and a pressure of 101 kPa (1 atm) and the concentration of any solution is 1 mol dm^{-3}.

Standard molar enthalpy change of combustion

→ $\Delta H^{\ominus}_{c,298}$ represents the heat change at *constant pressure* when one mole of the substance (at 298 K and 1 atm) is *completely* burnt to form products (at 298 K and 1 atm).

Enthalpies of combustion are particularly important for organic compounds and may be shown with an equation. For example:

$$C_4H_9NH_2(l) + 6\tfrac{3}{4}O_2(g)$$
$$\rightarrow 4CO_2(g) + 5\tfrac{1}{2}H_2O(l) + \tfrac{1}{2}N_2(g); \Delta H^{\ominus}_{c,298} = -3\,018 \text{ kJ mol}^{-1}$$

The 3 018 kJ of heat released to the surroundings includes $5\tfrac{1}{2} \times 41$ kJ of heat released when the water condenses to a liquid because

$$H_2O(g) \rightarrow H_2O(l); \Delta H^{\ominus} = -41 \text{ kJ mol}^{-1}$$

Watch out!

The complete combustion of nitrogen-containing organic compounds does not produce oxides of nitrogen.

Standard molar enthalpy change of formation

→ $\Delta H^{\ominus}_{f,298}$ represents the heat change at *constant pressure* when one mole of the substance (at 298 K and 1 atm) is formed from its constituent elements in their most stable form at 298 K and 1 atm.

The enthalpy change of combustion of some elements is the same as the enthalpy change of formation of their oxides. For example:

$$C(\text{graphite}) + O_2(g) \rightarrow CO_2(g); \Delta H^{\ominus}_{298} = -393.5 \text{ kJ mol}^{-1}$$

Don't forget!

Many compounds cannot be formed by direct combination of their elements. In those cases $\Delta H^{\ominus}_{f,298}$ must be found indirectly using Hess's law.

Hess's law ●●●

According to the first law of thermodynamics

→ energy cannot be created and cannot be destroyed

Germain Henri Hess stated his law of constant heat summation in 1840. It is a particular case of the more general first law of thermodynamics. **Hess's law** has been expressed in many ways:

→ the standard molar enthalpy change of a process is independent of the means or route by which that process takes place.

Watch out!

Strictly speaking, Hess's law breaks down if, as well as enthalpy changes, the processes involve other forms of energy changes.

Exam question (4 min)
answer: page 39

Use the following standard enthalpy changes of combustion to determine the standard enthalpy change of formation of methane. Include the sign in your answer and state whether methane is an exothermic or endothermic compound.

Substance	Standard enthalpy change of combustion/kJ mol^{-1}
Hydrogen, $H_2(g)$	−286
Carbon, C(graphite)	−394
Methane, $CH_4(g)$	−890

Examiner's secrets

You should know how to use Hess's law to calculate enthalpy changes that cannot be determined by direct experimental measurement. You will get marks for drawing a clearly labelled energy cycle diagram.

Bond energies

When a covalent bond forms between two atoms, energy is given out and the process is exothermic. When a covalent bond between two atoms is broken, energy is taken in and the process is endothermic. The amount of energy involved in these processes depends upon the nature of the atoms and the bonds that join them.

Bond dissociation enthalpies ○○○

The difference in the standard molar enthalpies of combustion of the lower alkanes and of the primary alcohols is about 650 kJ mol^{-1}. The difference in molecular formula from one alkane (or alcohol) to the next is CH_2. So the 650 kJ corresponds to the combustion of a CH_2 group to form CO_2 and H_2O.

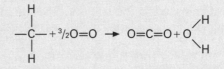

We suppose the 650 kJ represents the difference between the energy taken in by the breaking of the C—H, O=O bonds and the energy given out by the forming of the C=O and O—H bonds. It follows that bonds need definite energies to break them. Here are the first and second standard bond dissociation enthalpies of water:

first $H-OH(g) \rightarrow H(g) + OH(g);$ $\Delta H^{\oplus} = +498 \text{ kJ mol}^{-1}$

second $O-H(g) \rightarrow O(g) + H(g);$ $\Delta H^{\oplus} = +430 \text{ kJ mol}^{-1}$

→ The precise strength of a bond (X – Y) depends upon the other atoms or groups of atoms attached to X and Y.

Bond energy ○○○

In a data book the H—O bond energy is listed as $E(H-O) = +464 \text{ kJ mol}^{-1}$ which is the average value of the first and second standard bond dissociation enthalpies of water. $E(C-H) = 435 \text{ kJ mol}^{-1}$ is an average value of the standard bond dissociation enthalpies of the four bonds in methane: namely CH_3-H, CH_2-H, $CH-H$, $C-H$.

→ Bond energy is the *average* standard enthalpy change for the breaking of a mole of bonds in a gaseous molecule to form gaseous atoms.
→ Bond energies indicate the strength of the forces holding together atoms in a covalent molecule.

Calculating bond energies with Hess's law ○○○

You should understand how to use standard molar enthalpy changes of formation of molecules and of atomization of elements to calculate an average standard molar bond enthalpy.

→ The standard molar enthalpy change of **atomization** of an element is the enthalpy change when *one mole of gaseous atoms* are formed from the element under standard conditions.

Example:
Calculating
$E(C{-}H)$ for
methane

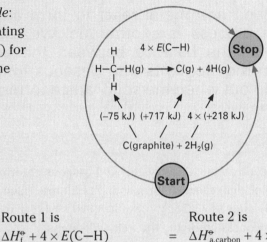

Route 1 is
$$\Delta H_f^\ominus + 4 \times E(C{-}H)$$
$$\Rightarrow (-75) + 4 \times E(C{-}H)$$
$$\Rightarrow E(C{-}H) = +416 \text{ kJ mol}^{-1}$$

Route 2 is
$$= \Delta H_{a,carbon}^\ominus + 4 \times \Delta H_{a,hydrogen}^\ominus$$
$$\quad (+717) \quad\; + 4 \times (+218)$$

Watch out!

For atomization of elements, molar refers to one mole of atoms produced: $^1/_2H_2(g) \to H(g)$, $\Delta H = +218$ kJ mol^{-1} It does *not* refer to one mole of element being atomized!
$H_2(g) \to 2H(g)$; $\Delta H = +436$ kJ mol^{-1}

Patterns in bond energies

→ Bond energies range from 150 kJ mol^{-1} for weak bonds (HO—OH, I—I, F—F, $H_2N{-}NH_2$) to 350–550 kJ mol^{-1} for strong bonds (C—C, C—H, N—H, O—H, H—Cl) to around 1 000 kJ mol^{-1} for very strong bonds (N≡N, C≡C, C≡O).

→ Bond energy increases with the number of shared electron-pairs ($E(C{-}C) = +347$, $E(C{=}C) = +612$, $E(C{\equiv}C) = +838$ kJ mol^{-1}).

Calculations using bond energies

You could be asked to calculate an approximate value for the enthalpy change of a reaction.

Example: Estimate the enthalpy change for the catalytic reforming of hexane to cyclohexene and hydrogen which is part of the process for the industrial production of petrol: $C_6H_{14} \to C_6H_{12} + H_2$.

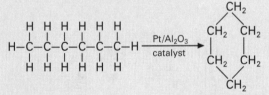

Break 2 C—H bonds
$\Rightarrow 2 \times 413$ kJ used up
$\Rightarrow 826$ kJ used up
$\Rightarrow$ net release of $1\,130 - 826 = 304$ kJ
$\Rightarrow \Delta H_f^\ominus = -304$ kJ mol^{-1}

Form 2 C—C bonds and
1 H—H bond
2×347 and 1×436 kJ given out
$1\,130$ kJ given out

Exam question (10 min)   answer: page 39

(a) Calculate the mean bond enthalpy $E(N{-}H)$ for ammonia. $\Delta H_f^\ominus$ for ammonia = −46.1 kJ mol^{-1} and the enthalpy changes of atomization of nitrogen and hydrogen are 473 and 218 kJ mol^{-1} respectively.

(b) How does the bond energy, bond length and thermal stability of the hydrogen halides vary with increasing molar mass from HF to HI?

Lattice energies

When a metal and non-metal react to form an ionic compound, the process is exothermic. We see this energy output as the balance between the energy used up to turn the elements into gaseous ions and the energy given out when the solid lattice forms from the gaseous ions.

Energy cycles

Solid sodium (body-centred metallic crystal) and gaseous chlorine (simple diatomic molecules) react to form sodium chloride (face-centred ionic crystal): $Na(s) + \frac{1}{2}Cl_2(g) \rightarrow NaCl(s)$; $\Delta H_f^\ominus = 411$ kJ mol^{-1}. We can imagine the process involving the following steps.

1 Sodium vaporizes and the gaseous atoms lose electrons:

$$Na(s) \rightarrow Na(g); \Delta H_{at}^\ominus = +107 \text{ kJ mol}^{-1}$$
$$Na(g) \rightarrow e^- + Na^+(g); \Delta H_i^\ominus = +496 \text{ kJ mol}^{-1}$$

2 Chlorine molecules dissociate and the gaseous atoms gain electrons:

$$\frac{1}{2}Cl_2(g) \rightarrow Cl(g); \Delta H_i^\ominus = +122 \text{ kJ mol}^{-1}$$
$$Cl(g) + e^- \rightarrow Cl^-(g); \Delta H_e^\ominus = -349 \text{ kJ mol}^{-1}$$

3 Gaseous sodium cations and chloride anions form a crystal lattice:

$$Na^+(g) + Cl^-(g) \rightarrow NaCl(s); \Delta H_l^\ominus$$

→ Lattice energy is the exothermic heat of formation of one mole of an ionic solid from its constituent *gaseous ions*.

We can draw an energy cycle for these steps and use Hess's law to calculate a value for the lattice energy. This is usually called a Born–Haber cycle.

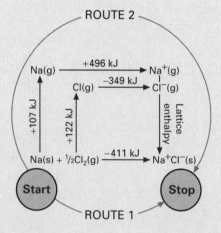

Route 1 = Route 2
$(-411) = (+107) + (+496) + (+122) + (-349) + \Delta H_e^\ominus$
$\Rightarrow \Delta H_e^\ominus = (-411) - (+107) - (+496) - (+122) - (-349)$
$\Rightarrow \qquad = -787 \text{ kJ mol}^{-1}$

Born–Haber cycles

It is instructive to present energy cycles like the **Born–Haber cycle** to show the energy changes against a vertical scale.

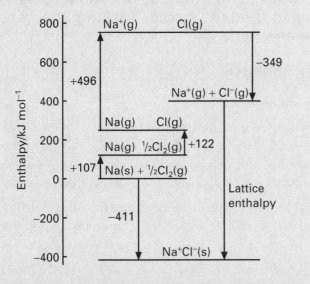

The jargon

The vertical scale represents the enthalpy H (or heat content). For elements in their most stable states, $H = 0$ by definition. Strictly speaking, the diagram does not need the arrow heads and the + or − signs. ↑ and + mean endothermic. ↓ and − mean exothermic.

Lattice energy values

You will not have to remember specific lattice energy values but you should have some idea of their range and magnitude compared to bond energies.

→ Lattice enthalpies range from around 600 kJ mol^{-1} to more than $4\,000 \text{ kJ mol}^{-1}$ which is well beyond the strongest covalent bond.

You should remember and be able to suggest explanations for certain patterns in the values of lattice energies.

→ Lattice energies for doubly charged ions are usually much more exothermic than those for singly charged ions.
→ Lattice energies become less exothermic as the size of the anion increases.
→ Lattice energies become less exothermic as the size of the cation increases.

You should appreciate how experimental and theoretical lattice energies indicate the character of alkali metal and silver halides.

→ Alkali metal halides are ionic with little or no covalent character.
→ Silver halides have a considerable percentage of covalent character.

Action point

Verify these patterns by looking up the lattice enthalpies for the following: LiF, LiCl, LiBr, LiI, NaF, NaCl, KF, RbF, CsF, Na_2O and $MgCl_2$.

The jargon

An *experimental lattice energy* is determined using a Born–Haber cycle like the one above for NaCl. A *theoretical lattice energy* is a mathematical prediction based on a model of perfectly spherical ions carrying complete charges.

Examiner's secrets

Remember that marks are given for a clearly labelled energy cycle.

Exam question (10 min)   answer: page 40

Ethene, C_2H_4, burns completely in oxygen to form carbon dioxide and water. Ethene also reacts with chlorine to form 1,2-dichloroethane, CH_2ClCH_2Cl. Calculate the enthalpy change of formation

(a) of ethene from its elements, and

(b) of 1,2-dichloroethane from ethene and chlorine.

Use the following molar enthalpies of combustion: $H_2(g)$ −286; $C(s)$ −394; $C_2H_4(g)$ −1 411, $CH_2ClCH_2Cl(l)$ −1 246 kJ mol^{-1}.

Free energy and entropy: what makes reactions go?

The jargon

Stable kinetically = very high E_a (activation energy) value.

Not feasible energetically = (free) energy of products > (free) energy of reactants.

A chemical reaction may not work for two reasons. On the one hand, the reactant mixture may be extremely stable kinetically, making the reaction infinitely slow. On the other hand, the reactant mixture may be energetically extremely stable, making the reaction not feasible energetically.

Energy changes ●●●

Enthalpy change, ΔH

→ Most reactions are exothermic (ΔH negative) and feasible because the products are energetically more stable than the reactants.

→ Most endothermic (ΔH positive) reactions are not feasible because the reactants are energetically more stable than the products.

You should be able to calculate the standard molar enthalpy change for a reaction, $\Delta H_r^{\ominus}$, from $\Delta H_f^{\ominus}$ values of the reactants and products. E.g.:

$$H^+(aq) + HCO_3^-(aq) \rightarrow H_2O(l) + CO_2(g)$$
$$\Delta H_f^{\ominus} \quad 0 \qquad -692 \qquad\qquad -286 \qquad -394 \text{ kJ mol}^{-1}$$

so $\Delta H_r^{\ominus}$ is $[(-286) + (-394)] - [(0) + (-692)] = +12 \text{ kJ mol}^{-1}$. This reaction of acid on hydrogencarbonate is endothermic and would not seem energetically feasible but it works!

→ ΔH is not always an accurate measure of energetic feasibility.

The jargon

G = Gibbs free energy after J. Willard Gibbs, Professor of Mathematical Physics at Yale University in 1875.

Free energy change, ΔG

ΔG is a thermodynamic quantity related to ΔH. You can find values for the standard free energy change of formation, $\Delta G_f^{\ominus}$, alongside $\Delta H_f^{\ominus}$ values in data books. And you could use them in the same way to calculate the standard free energy change, $\Delta G_r^{\ominus}$, for the reaction. For example:

$$H^+(aq) + HCO_3^-(aq) \rightarrow H_2O(l) + CO_2(g)$$
$$\Delta G_f^{\ominus} \quad 0 \qquad -587 \qquad\qquad -286 \qquad -394 \text{ kJ mol}^{-1}$$

so $\Delta G_r^{\ominus}$ is $[(-286) + (-394)] - [(0) + (-587)] = -93 \text{ kJ mol}^{-1}$.

→ A reaction will be energetically feasible (spontaneous) if its standard molar free energy change, $\Delta G_r^{\ominus}$, is negative.

Checkpoint

Calculate $\Delta H^{\ominus}$ and $\Delta G^{\ominus}$ for the process KI(s) → KI(aq) from the following data and comment on the feasibility of dissolving potassium iodide in water.

	$\Delta H_f^{\ominus}$	$\Delta G_f^{\ominus}$
KI(s)	-328 kJ mol^{-1}	-325 kJ mol^{-1}
KI(aq)	-307 kJ mol^{-1}	-335 kJ mol^{-1}

"The total entropy always increases if a spontaneous reaction occurs"

Second Law of Thermodynamics

Entropy changes ●●●

Free energy change ΔG is closely related to enthalpy change ΔH, and both are related to entropy change ΔS by the equation $\Delta G = \Delta H - T \times \Delta S$ where ΔG, ΔH and ΔS represent the *values at temperature T* of the changes in free energy, enthalpy and entropy.

→ $\Delta G^{\ominus} = \Delta H_f^{\ominus} - T \times \Delta S^{\ominus}$ where T is the standard temperature and the values of the changes in free energy, enthalpy and entropy are the standard values at the standard temperature T.

We can rearrange the equation, $-(\Delta G^{\ominus}/T) = -(\Delta H^{\ominus}/T) + \Delta S^{\ominus}$, and then replace $-(\Delta G^{\ominus}/T)$ by $\Delta S_{total}^{\ominus}$ and $-(\Delta H^{\ominus}/T)$ by $\Delta S_{surroundings}^{\ominus}$ to give

$$\Delta S_{total}^{\ominus} = \Delta S_{surroundings}^{\ominus} + \Delta S_{system}^{\ominus}$$

Since $\Delta G^{\ominus}$ is negative for feasible reactions and $\Delta S_{total}^{\ominus} = -\Delta G^{\ominus}/T$ then

→ $\Delta S_{total}^{\ominus}$ must be positive for a reaction to be energetically feasible.

Conditions for feasible reactions ●●●

→ A reaction is feasible if $\Delta G^{\ominus}$ is negative ($\Delta S^{\ominus}_{total}$ is positive).

$\Delta H^{\ominus}$	$\Delta S^{\ominus}$	$\Delta G^{\ominus} = \Delta H^{\ominus} - T \times \Delta S^{\ominus}$	Feasible?
−	+	Negative	Yes – always
−	−	Negative if $|T \times \Delta S^{\ominus}| < |\Delta H^{\ominus}|$	Maybe
		Positive if $|T \times \Delta S^{\ominus}| > |\Delta H^{\ominus}|$	
+	+	Negative if $|T \times \Delta S^{\ominus}| > |\Delta H^{\ominus}|$	Maybe
		Positive if $|T \times \Delta S^{\ominus}| < |\Delta H^{\ominus}|$	
+	−	Positive	No – never

→ Reactions are feasible when the total entropy increases.
→ Entropy increases significantly when gases are produced.

What makes electrochemical reactions feasible?

→ Zinc displaces lead from its aqueous lead(II) cations and the feasible reaction is $Zn(s) + Pb^{2+}(aq) \rightarrow Zn^{2+}(aq) + Pb(s)$

For $\quad Zn(s)|Zn^{2+}(aq) \vdots Pb^{2+}(aq)|Pb(s) \quad E^{\ominus}_{cell} = +0.63$ V
$\qquad\qquad \xrightarrow{\quad\quad} \quad 2e^- \quad \xrightarrow{\quad\quad}$

but for $\quad Pb(s)|Pb^{2+}(aq) \vdots Zn^{2+}(aq)|Zn(s) \quad E^{\ominus}_{cell} = -0.63$ V
$\qquad\qquad \xleftarrow{\quad\quad} \quad 2e^- \quad \xleftarrow{\quad\quad}$

→ $\Delta G^{\ominus} = -nFE^{\ominus}_{cell} = -2 \times 96\,500 \times (-0.63) = 121\,590$ kJ mol^{-1}.
→ Positive $E^{\ominus}_{cell}$ indicates the direction of the feasible cell reaction.

What makes reduction of oxides by carbon feasible?
An Ellingham diagram shows $\Delta G^{\ominus}_{f}$ of oxides against T.

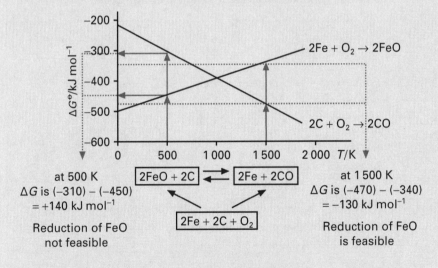

at 500 K
ΔG is $(-310) - (-450)$
$= +140$ kJ mol^{-1}

Reduction of FeO
not feasible

$2FeO + 2C \rightleftharpoons 2Fe + 2CO$

$2Fe + 2C + O_2$

at 1 500 K
ΔG is $(-470) - (-340)$
$= -130$ kJ mol^{-1}

Reduction of FeO
is feasible

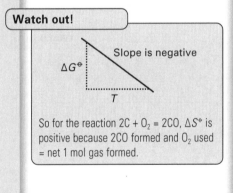

Exam question (10 min)   answer: page 40

Discuss the following in terms of changes in free energy and entropy.

(a) The reaction of sodium hydrogencarbonate with acid is endothermic but feasible.

(b) The reduction $Al_2O_3 + 3C \rightarrow 2Al + 3CO$ is feasible above 2100 K but the reduction $Al_2O_3 + 3CO \rightarrow 2Al + 3CO_2$ is not feasible at any temperature.

Structured exam question

answer: page 40

(a) The standard molar enthalpy change of combustion ($\Delta H^\ominus_{c,298}$) of methane is −890.3 kJ mol^{-1}. A value for the enthalpy change of combustion of methane may be calculated from standard molar enthalpy changes of formation ($\Delta H^\ominus_{f,298}$) or from bond dissociation energies (E).

(i) Calculate the enthalpy change of combustion of methane using the following values for the standard molar enthalpy changes of formation.

Compound	$\Delta H^\ominus_{f,298}$/kJ mol^{-1}
Carbon dioxide	−394
Water	−286
Methane	−74.8

....
....
....

(ii) Calculate the enthalpy change of combustion of methane using the following values for the average bond dissociation energies.

Bond	E/kJ mol^{-1}
C−H	+435
O=O	+498
C=O	+805
H−O	+464

....
....
....

(iii) Comment upon the values calculated in (i) and (ii) above.

....
....
....

(b) Calculate the enthalpy change of solution of solid sodium chloride from the following data.

Change	Enthalpy change/kJ mol^{-1}
$NaCl(s) \rightarrow Na^+(g) + Cl^-(g)$	+787
$Cl^-(g) + aq \rightarrow Cl^-(aq)$	−377
$Na^+(g) + aq \rightarrow Na^+(aq)$	−406

....
....
....
....
....
....
....
....
....
....
....

(10 min)

Examiner's secrets

You will impress if you keep the − and + signs with the values in brackets until your final step. Lose signs and you lose marks.

Watch out!

Don't forget that breaking bonds uses energy and therefore is endothermic with ΔH being positive.

Examiner's secrets

You may be given lines for writing or space for diagrams and/or calculations. The number of lines and the amount of space is a clue to how much is expected for a good answer *but* the number of marks (1 mark = 1 exam minute) is a better guide.

Answers
Energetics

Measuring enthalpy changes

Checkpoints

1 Heat produced is $100 \times 4.2 \times 6.9 = 2\,898$ J. Amount of acid was $1.0 \times 50/1\,000 = 0.05$ mol HCl(aq) and this reacted with the same amount of NaOH(aq). So the heat produced by 1 mol HC(aq) or NaOH(aq) would be $2\,898/0.05 = 58\,000$ Js. Hence the molar enthalpy change is -58 kJ mol^{-1}

2 (a) To pump
  (b) Thermometer
  (c) Stirrer to ensure a uniform temperature
  (d) Air (or oxygen)
3 J K^{-1} or kJ K^{-1}
4 Production of carbon dioxide adding to the greenhouse effect and global warming. Air pollution by carbon monoxide produced by incomplete combustion.

Exam question

(a) (i) Amount (in moles) of propan-2-ol used is $0.149/60.1 = 2.48 \times 10^{-3}$ mol. Heat produced is $2.48 \times 10^{-3} \times 2\,006$ kJ ($= 4.975$ kJ). This causes a temperature rise of 5.42 °C and so the heat to cause a rise of 1 °C is $(2.48 \times 10^{-3} \times 2\,006)/5.42 = 0.918$ kJ K^{-1}, the heat capacity of the calorimeter.
  (ii) Heat produced by 0.152 g of butan-1-ol is 0.918 kJ K$^{-1} \times 5.21$ K $= 4.78$ kJ.
  Heat produced by one mole ($= 74.1$ g) of butan-1-ol is $4.78 \times 74.1/0.152 = 2\,330$ kJ mol^{-1}.
(b) The student's value is too low. $2\,676 - 2\,330 = 346$ kJ of heat may have been 'lost' to the surroundings or not produced because combustion was incomplete. Surround the apparatus with thermal insulation. Enrich the air with oxygen to ensure more complete combustion. Use a more accurate thermometer. (Marks would be given for any sensible suggestion.)

ΔH and Hess's law

Checkpoints

1 (a) Oxygen at high pressure
  (b) Electrical connection to ignite sample
  (c) Stirrer to ensure a uniform temperature
  (d) Water to take up the heat produced
2 M_r(octane) = 114
  0.114 g is 0.01 mol
  Heat produced = 5.47 kJ
  $\Delta H_c = -5.47 \div 0.001 = -5.470$ kJ mol^{-1}

Exam question

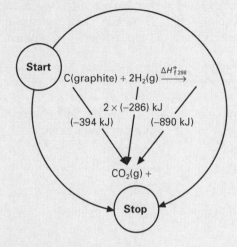

$\Delta H + (-890) = (-394) + (2 \times (-286))$
$\Delta H = +890 - 394 - 572$
$\quad = -76$ kJ mol^{-1}, an exothermic compound

Bond energies

Exam question

(a) Atomization enthalpy of hydrogen is $+218$ kJ mol^{-1}.

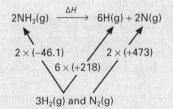

From the cycle

$\quad 2 \times (-46.1) + \Delta H = 6 \times (+218) + 2 \times (+473)$

so

$\quad -92.2 + \Delta H = 1\,308 + 946$

Hence $\Delta H = 2\,346.2$ kJ. This is for breaking six moles of N—H bonds. Therefore E(N—H) is $2\,346/6 = 391$ kJ mol^{-1}.
(b) From HF to HI the bond energy decreases and therefore the thermal stability decreases. Bond length increases from HF to HI.

Lattice energies

Checkpoint

(a) Adding an electron to the O$^-$ ion uses energy to overcome the repulsion between the negative charges so the enthalpy change for O$^-$(g) + e$^- \to$ O^{2-}(g) must be positive.
(b) The electron affinities become less exothermic from Cl to I. The nuclear charge increases from Cl to I but its attraction for the added electron does not increase

because the inner-shell shielding also increases from Cl to I. The increase in shielding outweighs the increase in nuclear charge so the attraction for the added electron decreases from Cl to I and less energy is released.

Exam question

(a)

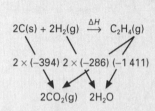

$\Delta H + (-1\,411) = 2 \times (-394) + 2 \times (-286)$
$\Delta H = -(-1\,411) + 2 \times (-394) + 2 \times (-286)$
$\Delta H = +51.0 \text{ kJ mol}^{-1}$

(b)

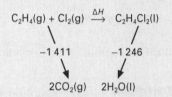

$\Delta H + (-1\,264) = -1\,411$
$\Delta H = -165 \text{ kJ mol}^{-1}$

Free energy and entropy: what makes reactions go?

Checkpoint

ΔH is $(-307) - (-328) = +21 \text{ kJ mol}^{-1}$
ΔG is $(-335) - (-325) = -10 \text{ kJ mol}^{-1}$
Potassium iodide will dissolve endothermically and spontaneously in water.

Exam question

(a) $\Delta G^{\ominus}$ is negative because $\Delta S^{\ominus}$ is highly positive when the solid forms a solution and gives off CO_2 gas. These products are more disordered than the reactants and the energy distribution more varied.

(b) Above 2 100 K, $\Delta G^{\ominus}$ for $3C + 1\frac{1}{2}O_2 \rightarrow 3CO$ is more negative ($\Delta S^{\ominus}$ positive) and therefore more feasible than $2Al + 1\frac{1}{2}O_2 \rightarrow Al_2O_3$. The formation of Al_2O_3 is always more negative and therefore more feasible than $3CO + 1\frac{1}{2}O_2 \rightarrow 3CO_2 (\Delta S^{\ominus}$ negative) at any temperature.

Structured exam question

(a) (i) Using an energy cycle:

$$CH_4 + 2O_2 \xrightarrow{\Delta H_c^{\ominus}} CO_2 + 2H_2O$$

$-74.8 \quad -394 \quad 2 \times (-286)$
Elements

$-74.8 + \Delta H_c^{\ominus} = (-394) + 2 \times (-286)$
Hence $\Delta H_c^{\ominus} = -891 \text{ kJ mol}^{-1}$ 3 sig fig

(ii) $CH_4 + 2O_2 \rightarrow CO_2 + 2H_2O$
$4 \times (+435) + 2 \times (+498) \quad 2 \times (+805) + 2 \times [2 \times (+464)]$
So $\Delta H_c^{\ominus}$ is
$-\{2 \times (+805) + 2 \times [2 \times (+464)] - [4 \times (+435)$
$+ 2 \times (+498)]\} = -\{[1\,610 + 1\,856] - [1\,740 + 996]\}$
$= -730 \text{ kJ mol}^{-1}$

(iii) The value $-891.2 \text{ kJ mol}^{-1}$ should be very close to the experimental value because the standard molar enthalpies of formation are obtained indirectly from experimentally determined enthalpies of combustion. The value -730 kJ mol^{-1} could be expected to differ from $-891.2 \text{ kJ mol}^{-1}$ because the bond dissociation energies are average energy values and may be determined by interpretation of data such as those obtained from spectroscopy.

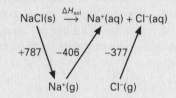

$\Delta H_{sol} = (+787) + (-406) + (-377)$
So $\Delta H_{sol} = +4 \text{ kJ mol}$

Kinetics

You probably know how to make hydrochloric acid react faster with magnesium: use more concentrated acid, use magnesium powder instead of ribbon and warm the mixture. In this chapter you will learn more about rates of reactions and how they depend upon concentration and temperature. You will see how collision theory explains this dependence. You will also discover how catalysts work. Make sure you understand the maths behind an expression like 'rate = $k[A]^m[B]^n$'; it's easier than it looks.

Exam themes

→ Describe a way of following the rate of a specified reaction

→ Describe how and why temperature, concentration and catalyst affect the rate of a reaction and use the concept of activation energy and the Maxwell–Boltzmann distribution to explain the effects

→ Describe the way in which homogeneous and heterogeneous catalysts work and give examples of homogeneous and heterogeneous catalysts in inorganic and organic chemistry

→ Describe what is meant by a rate equation and, using supplied data, work out the order of reaction, the rate equation, the rate constant and its units

→ Work out a possible mechanism from a rate equation or select which of several supplied mechanisms best fits the rate equation

→ Draw and interpret energy level diagrams (reaction profiles) to describe uncatalysed and catalysed reactions

Topic checklist

O AS ● A2	AQA	CCEA	EDEXCEL	OCR	WJEC
Rates and orders of reaction	O	O	O	O	O
Orders of reaction and mechanisms	●	●	●	●	●
Reaction rates and temperature	●	●	●	●	●
Catalysis	●	●	●	●	●

Rates and orders of reaction

During any chemical reaction the concentrations of reactants decrease and the concentrations of products increase. Rates of reaction are measures of how these concentrations change with time.

Rate of reaction

●●●

The rate of a reaction is the change of concentration of a reactant (or of a product) with time.

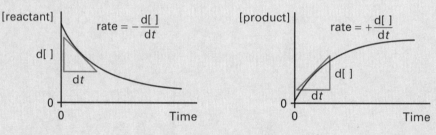

Units of rate are usually (but not always) $mol\,dm^{-3}\,s^{-1}$.

Measuring rates of reaction

We follow reactions by measuring the concentration of a reactant (or product) or by monitoring changes in a property of the system. The following are important reactions for A-level chemistry. They proceed at a rate that you could conveniently measure in your laboratory.

Example of reaction	Property measured
$Mg(s) + 2HC(aq) \rightarrow MgCl_2(aq) + H_2(g)$	Gas volume
$CaCO_3(s) + 2HCl(aq) \rightarrow CaCl_2(aq) + H_2O(l) + CO_2(g)$	Mass
$CH_3CO_2C_2H_5(l) + H_2O(l) \rightarrow CH_3CO_2H(aq) + C_2H_5OH(aq)$	Concentration
$CH_3COCH_3(aq) + I_2(aq) \rightarrow CH_3COCH_2I(aq) + HI(aq)$	Colour intensity

Experiment 1

Experiment 2

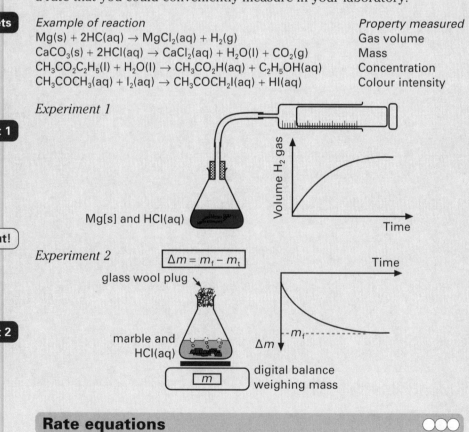

Rate equations

$$rate = k[A]^m[B]^n$$

→ k is the **rate constant** whose value depends upon temperature.
→ m and n are the **orders of reaction** with respect to the reactants.
→ $m + n$ is called the **overall order** of the reaction.

The jargon

$[A]$ = concentration of A where A is a reactant or a product. $d[\,]/dt$ is the slope (gradient or tangent) at a point on a graph of concentration $[\,]$ against time t and is related to the rate of reaction. Rate $= d[A]/dt$ when A is a product but rate $= -d[A]/dt$ when A is a reactant. The negative sign in front makes the rate values positive.

Examiner's secrets

You should know the details of simple experiments you yourself could do to measure reaction rates with standard laboratory apparatus.

Checkpoint 1

(a) State and explain three precautions needed in experiment 1.
(b) Suggest how you could start the reaction and the clock *both together*.

Watch out!

Make sure you understand why the direction of the y-axis in experiment 1 is the opposite of the y-axis in experiment 2.

Checkpoint 2

(a) State and explain the purpose of the glass wool plug in experiment 2.
(b) If m_t is the reading on the digital balance at time t and m_f is the final reading when the reaction has stopped, what is Δm proportional to?

For A-level chemistry the orders m and n may be 0, 1 or 2 but the overall order $m + n$ will not exceed 2. The associated integrated rate equations express concentration as a function of time.

Zero order reactions

$$-d[\ ]/dt = k_0[\text{reactant}]^0 \quad \text{and} \quad [\text{reactant}] = -k_0t + [\text{reactant}]_{\text{initial}}$$

→ Rate is independent of concentration.
→ The value of k_0 may be obtained from the slope of a graph of concentration against time.
→ The units of k_0 are $\text{mol dm}^{-3}\text{s}^{-1}$.
→ Half-life = $[\ ]/2k_0$.

First order reactions

$$-d[\ ]/dt = k_1[\text{reactant}]^1 \quad \text{and} \quad \ln[\text{reactant}] = -k_1t + \ln[\text{reactant}]_{\text{initial}}$$

→ Half-life = $(\ln 2)/k_1$ is independent of concentration.
→ The value for k_1 may be obtained from the slope of a graph of natural logarithm of concentration against time or from the half-life.
→ The units of k_1 are s^{-1}.

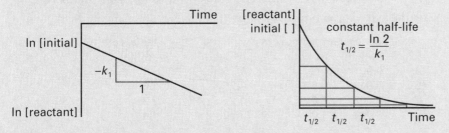

Second order reactions

$$-d[\ ]/dt = k_2[\text{reactant}]^2 \quad \text{and} \quad 1/[\text{reactant}] = k_1t + 1/[\text{reactant}]_{\text{initial}}$$

→ The value for k_2 may be obtained from the slope of a graph of 1/concentration against time.
→ The units of k_2 are $\text{mol}^{-1}\text{dm}^3\text{s}^{-1}$.
→ Half-life = $1/(k_2[\ ])$.

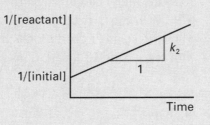

> **Watch out!**
>
> The value of k *does not* depend upon concentration. k depends only upon the temperature and the reaction chosen.

> **The jargon**
>
> *Half-life* is the time interval for the concentration of a reactant to decrease to half its original value.

> **The jargon**
>
> ln means logarithm to the base e. log means logarithm to the base 10.

> **Don't forget!**
>
> The decay and half-life of radioactive isotopes provide the best examples of first order reactions.

> **Don't forget!**
>
> The *amount* of product formed in any reaction always depends upon the *amount* of reactants even though the rate of reaction may be independent of the concentration of a particular reactant. Reactants always react to produce products!

Exam question (5 min)   answer: page 51

Carbon-14 has a half-life of 5730 years and decays to nitrogen-14.

(a) Write an equation for the decay of carbon-14.
(b) How long would it take for the radioactivity of a carbon-14 sample to fall to 1/8 of its original value?
(c) Why is carbon-14 important to archaeologists?

Orders of reaction and mechanisms

All rate equations must be determined experimentally; they cannot be predicted from the stoichiometric equation for a reaction. Consequently we must do experiments to find the orders and the value of the rate constant for any reaction. The rate equation may reveal the mechanism of the reaction.

The jargon

A *stoichiometric equation* is a balanced chemical equation showing the relative amounts of reactants and products. The (usually simple whole) numbers in front of each formula are known as stoichiometric coefficients.

Finding the orders of a reaction ●●●

The experimentally determined stoichiometry of the acid-catalysed reaction of propanone and iodine is given by the following equation:

$$CH_3COCH_3(aq) + I_2(aq) \rightarrow CH_3COCH_2I(aq) + H^+(aq) + I^-(aq)$$

To find the orders of reaction with respect to propanone, iodine and acid catalyst we must do several experiments to see how the concentration of each substance alters the rate.

Checkpoint 1

State and explain two different ways of following the reaction of propanone with iodine.

Using initial rates

If you measure the *initial* rate at the start of the reaction (at the moment of mixing the reactants when $t = 0$) then you will also know the *initial* concentrations of all the substances. So if you do several separate experiments doubling the concentration of each substance in turn, you should discover how the concentration of each affects the rate. Here are some typical results:

Watch out!

Drawing tangents to a [ ] vs t curve, especially at the start where $t = 0$, can be very inaccurate.

	Initial concentrations/mol dm^{-3}			Initial rate/10^{-5} mol dm^{-3} s^{-1}
Expt	[iodine]	[propanone]	[HCl(aq)]	$-d[I_2(aq)]/dt$
A	0.001	0.5	1.25	1.1
B	0.002	0.5	1.25	1.1
C	0.002	1.0	1.25	2.2
D	0.002	1.0	2.50	4.4

Examiner's secrets

There are often questions asking you to find the orders of a reaction from a table of experimental data.

Experiments A and B show that the iodine concentration does *not* affect the rate, so the reaction is *zero order* with respect to $[I_2(aq)]$. Experiments B and C show that the rate doubles when the propanone concentration is doubled, so the reaction is *first order* with respect to $[CH_3COCH_3(aq)]$. Experiments C and D show that the rate doubles when the acid concentration is doubled, so the reaction is *first order* with respect to $[HCl(aq)]$. Consequently the rate equation is

$$-d[I_2(aq)]/dt = k[I_2(aq)]^0[CH_3COCH_3(aq)]^1[HCl(aq)]^1$$

or simply

$$rate = k[CH_3COCH_3(aq)][HCl(aq)]$$

Using half-lives

The variation in half-life with concentration (or time) may indicate the order of a reaction, e.g. for the reaction $N_2O_5(g) \rightarrow N_2O_4(g) + \frac{1}{2}O_2(g)$

Checkpoint 2

For the decomposition of $N_2O_5(g)$
(a) write a rate equation
(b) state the order of the reaction and
(c) plot a graph of ln $[N_2O_5]$ vs time to find a value for k

Approx [ ] values halve									
	180	90				30	15	10	5
$[N_2O_5(g)]/10^{-4}$ mol dm^{-3}	176	125	93	71	53	39	29	16	9	5
Time/10^2 s	0	6	12	18	24	30	36	48	60	72

Half-life approx 1 200 s   12   12   12

Reaction mechanisms

A reaction mechanism is a set of theoretical steps proposed to account for the conversion of reactants into products. For A-level chemistry, a reaction mechanism is usually considered to be a sequence of simple steps, the slowest of which controls the overall rate of the reaction.

→ The slowest step is called **the rate-determining step**.

Iodination of propanone

Here is one plausible mechanism for the propanone–iodine reaction.

$$(CH_3)_2CO(aq) + H_3O^+(aq) \xrightarrow{\text{slow}} (CH_3)_2COH^+(aq) + H_2O(l)$$

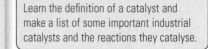

Each step is bimolecular. The (slow) rate-determining step involves one propanone molecule and one hydrogen ion to be consistent with the orders in the rate equation. The second step regenerates the H_3O^+ ion used up in the first step to be consistent with the hydrogen ion being a catalyst. When you combine the equations for the three steps (by adding the left-hand sides together, the right-hand sides together and cancelling identical species on both sides) you get the correct stoichiometric equation for the overall chemical reaction.

Hydrolysis of bromoalkanes

$$C_4H_9Br(l) + H_2O(l) \rightarrow C_4H_9OH(l) + HBr(aq)$$

You could do simple test-tube reactions to show that the hydrolysis of 2-bromo-2-methylpropane is much faster than that of 1-bromobutane.

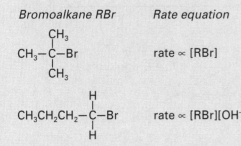

Bromoalkane RBr	Rate equation	Mechanism
	rate ∝ [RBr]	S_N1 (a two-step mechanism) $RBr \xrightarrow{\text{slow}} R^+$ $R^+ + OH^- \xrightarrow{\text{fast}} ROH$
	rate ∝ [RBr][OH⁻]	S_N2 (a one-step mechanism) $RBr + OH^- \xrightarrow{\text{slow}} ROH + Br^-$

Links

See page page 143: hydrohalogenation of alkenes; 144: halogenation of alkanes; page 145: halogenation of alkenes; page 147: nitration of arenes.

The jargon

Bimolecular refers to a step involving *two* reacting species. *Unimolecular* refers to a step involving *one* reacting species only.

Action point

Learn the definition of a catalyst and make a list of some important industrial catalysts and the reactions they catalyse.

Checkpoint 3

Combine the equations for the three steps and write down the resulting stoichiometric equation.

Watch out!

Don't confuse the meanings of S_N1 and S_N2. 1 stands for unimolecular. 2 stands for bimolecular. The 1 and 2 do *not* refer to the number of steps in the mechanism.

Exam question (5 min)   answer: page 51

(a) What is the overall order of the iodine and propanone reaction?

(b) Use the experimental data on page 44 to calculate a value for k.

(c) What are the units of k?

Reaction rates and temperature

Chemical reactions get faster as temperature increases. In many cases the rate of a reaction (and therefore the value of its rate constant, k) approximately doubles with every 10° rise in temperature. This effect of temperature upon reaction rates may be explained by the theory of effective collisions and activation energy.

Measuring the effect of temperature upon rates ●●●

When we add hydrochloric acid to aqueous sodium thiosulphate the clear, colourless solution slowly turns milky white as colloidal sulphur forms. The stoichiometric equation is: $H_2S_2O_3(aq) \rightarrow H_2SO_3(aq) + S(s)$. When we investigate the kinetics we find that the decomposition is a first order reaction and rate = $k_1[H_2S_2O_3(aq)]$. When we perform the same experiment at different temperatures and measure the time interval Δt to the same degree of milkiness, we find Δt approximately halves with each 10° rise in temperature.

Activation energy ●●●

If we increase only the temperature and keep all other conditions constant, the reaction rate increases and so the value of the rate constant k must increase.

→ The Arrhenius equation, $k = Ae^{-E_a/RT}$, expresses mathematically the dependence of the rate constant upon the absolute temperature.
→ E_a is called the activation energy of the reaction.

The Arrhenius equation may be written in the form $\ln k = \ln A - E_a/RT$ so that a value for E_a may be determined from the gradient (= $-E_a/R$) of a graph of $\ln k$ against $1/T$.

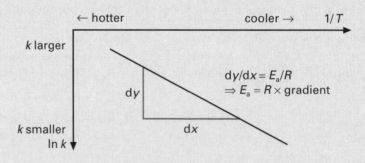

Theory of effective collisions

Billions of close encounters between particles (atoms, ions and molecules) occur every second but only a small number occur as collisions with enough energy (E_a) to activate a reaction.

→ The Arrhenius factor A is a measure the billions of close encounters occurring every second between particles.
→ $e^{-E_a/RT}$ is a measure of the (small) fraction of close encounters that result in collisions causing a reaction.
→ The lower the activation energy and the higher the temperature the faster the reaction.

You may consider activation energy from two points of view.

E_a as the energy barrier

The activation energy may be regarded as the *minimum energy* needed by the reacting particles (atoms, ions or molecules) to achieve a transition state where they exist as an activated complex. You should understand and be able to represent this idea in energy diagrams.

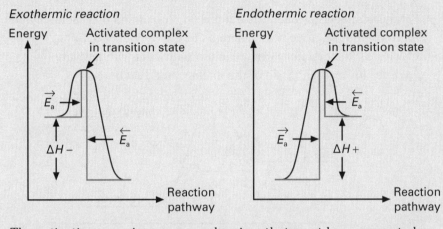

The activation energies appear as barriers that must be surmounted before reactants can turn into products and before products can re-form into reactants.

→ Breaking of strong covalent bonds requires energy and may lead to high activation energies resulting in slow reactions.

E_a and the distribution of energy

As a factor in the Arrhenius equation, E_a determines the proportion of reacting species (atoms, ions or molecules) having enough energy to cross the activation energy barrier. You should understand and be able to represent this idea by sketching energy distribution diagrams. Note that as the temperature rises

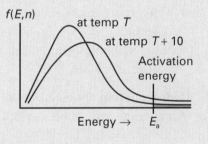

→ the peak of the curve moves *to the right*, so the mean value of $f(E,n)$ and therefore the mean energy of the molecules increases
→ the curve *flattens*, so the total area under it and therefore the total number of molecules remains constant
→ the area under the curve to the right of E_a, and therefore the number of molecules colliding with enough energy to cause a reaction, roughly doubles for every 10° rise in temperature.

The jargon

These energy diagrams may be called *reaction profiles*. The activated complex may be regarded as a transition state in which old bonds have partly broken and new bonds have partly formed.

Checkpoint 3

(a) How is the enthalpy change ΔH for a reaction related to the forward and reverse activation energies?
(b) Explain why the exothermic direction of a reaction is kinetically more feasible than its reverse endothermic direction?

The jargon

In these so-called *Maxwell–Boltzmann distributions of energies*, $f(E,n)$ is a function of energy such that the total area under the curve of $f(E,n)$ against energy is proportional to the total number of reactant species.

Watch out!

Energetically speaking, reactions with very large (positive or negative) values of ΔH may be considered irreversible. Kinetically speaking, reactions with very large values of E_a may be too slow to occur. Do not confuse energetic stability and kinetic stability!

Exam question (5 min)   answer: page 51

Sketch a reaction profile for $H_2(g) + I_2(g) \rightarrow 2HI(g)$ and comment on the reversibility of the reaction. [$\Delta H = -10$ kJ mol^{-1}; $E_a = 151$ kJ mol^{-1}]

Catalysis

So what about those really slow reactions? How can we speed them up? Higher temperature, higher concentrations, but don't forget the catalyst. Homogeneous, heterogeneous and biological: they can all speed up chemical reactions.

General features of catalysts

→ Catalysts speed up reactions but are not consumed by reactions and therefore do not appear as reactants in the overall equations.
→ Catalysts provide alternative reaction pathways with activation energies lower than those of the uncatalysed reactions.

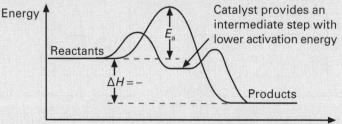

→ Catalysts speed up the rate of attainment of equilibrium for a reversible reaction without altering its composition at equilibrium.
→ Catalysts lower the activation energies of the forward and reverse reactions of a reversible reaction by the same amount.

Heterogeneous catalysis

→ A heterogeneous catalyst is a catalyst in a *different phase* from the reactants.
→ Many industrially important heterogeneous catalysts are d-block transition metals.

Hydrogenation of unsaturated oils to saturated fats in the production of margarine is a simple example. Nickel powder is the catalyst.

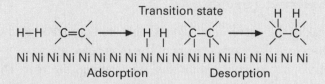

Homogeneous catalysis

→ A homogeneous catalyst is a catalyst in the *same phase* as the reactants.
→ Homogeneous catalysts take part in a reaction so an increase in their concentration will speed up the rate-determining step.

The oxidation of iodide anions by peroxodisulphate(VI) anions is slow: $S_2O_8^{2-}(aq) + 2I^-(aq) \rightarrow 2SO_4^{2-}(aq) + I_2(aq)$. However, iron(II) or iron(III) cations catalyse the reaction by taking part and providing an alternative pathway with a lower overall activation energy.

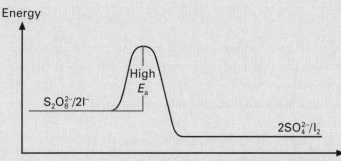

In the catalysed reaction, iron(II) cations reduce peroxodisulphate anions and iron(III) cations oxidize iodide anions:

$$2Fe^{2+}(aq) + S_2O_8^{2-}(aq) \rightarrow 2Fe^{3+}(aq) + 2SO_4^{2-}(aq)$$
$$2Fe^{3+}(aq) + 2I^-(aq) \rightarrow 2Fe^{2+}(aq) + I_2(aq)$$

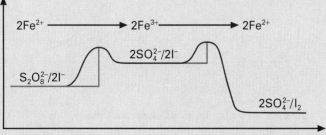

Checkpoint 2

Why might we expect the reaction between two anions to have a high activation energy and be slow?

Autocatalysis ●●●

In most reactions we expect the concentration of catalyst to remain constant. In the iodination of propanone, the concentration of the homogeneous catalyst, $H^+(aq)$, actually increases because hydriodic acid is one of the products. Consequently, the rate of the reaction at first increases with time even though the concentration of propanone is decreasing.

→ A reaction is autocatalytic if it is catalysed by one of its products.

Exam questions (10 min) answers: page 52

1 (a) Suggest why E_a for the reaction of Fe^{2+} with the peroxodisulphate ion might be larger than E_a for the reaction of Fe^{3+} with the iodide ion.
   (b) State the effect, if any, of the catalyst upon the value of ΔH.
   (c) Redraw the catalysed reaction profile to show iron(III) cations acting as the catalyst.

2 The ethanedioate ion reacts with the manganate(VII) ion in excess aqueous acid to form carbon dioxide and the manganese(II) ion. Write a balanced ionic equation for the redox and suggest *two* reasons why the reaction accelerates initially.

49

Structured exam question

answers: page 52

(a) The diagram below shows the distribution of molecular velocities in a given mass of gas at a particular temperature. The vertical axis is a function of the velocity.

Draw on the grid a curve showing the distribution of molecular velocities

(i) at a lower temperature – label this curve L

(ii) at a higher temperature – label this curve H

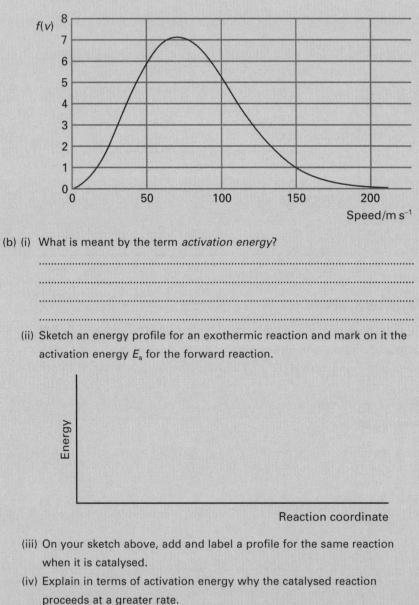

(b) (i) What is meant by the term *activation energy*?

....
....
....
....

(ii) Sketch an energy profile for an exothermic reaction and mark on it the activation energy E_a for the forward reaction.

Energy

Reaction coordinate

(iii) On your sketch above, add and label a profile for the same reaction when it is catalysed.

(iv) Explain in terms of activation energy why the catalysed reaction proceeds at a greater rate.

....
....
....
....
....
....
....
....

(10 min)

Answers
Kinetics

Rates and orders of reaction

Checkpoints

1 (a) Magnesium metal must be clean. Any oxide coating would react first to give water instead of hydrogen gas. Care must be taken that the syringe does not stick. The volume measured would be incorrect. The amount of reactants must not produce a volume of hydrogen greater than the capacity of the syringe.
  (b) Put the acid in a test-tube inside the conical flask which can be tilted to allow the acid to make contact with the metal as the clock is started.
2 (a) To prevent loss of mass due to spray from the acid.
  (b) $\Delta m = (m_t - m_f) \propto [\text{HCl(aq)}]_t$, the concentration of unreacted hydrochloric acid at time t.

Exam question

(a) $^{14}_{6}\text{C} \rightarrow {}^{14}_{7}\text{N} + \beta$
(b) It would take three half-lives, i.e. 17 190 years.
(c) The isotope is the basis for carbon dating of relics.

Orders of reaction and mechanisms

Checkpoints

1 Colorimetry, because the intensity of the brown colour of aqueous iodine decreases with time as the I_2(aq) is converted into colourless I^-(aq).
  Sampling, because the unreacted iodine could be titrated with aqueous sodium thiosulphate using starch as an indicator for the end-point of the titration.
2 (a) Rate = $k[\text{N}_2\text{O}_5]$
  (b) First order (1)
  (c)

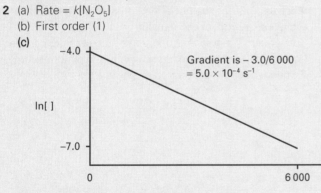

Gradient is $-3.0/6\,000$
$= 5.0 \times 10^{-4}\,\text{s}^{-1}$

Rate constant $k = 5.0 \times 10^{-4}\,\text{s}^{-1}$
3 $(\text{CH}_3)_2\text{CO(aq)} + I_2\text{(aq)} \rightarrow \text{CH}_3\text{COCH}_2\text{I(aq)} + \text{HI(aq)}$

Exam question

(a) The overall order is 2.
(b) According to the rate equation
$$\text{rate} = k[\text{CH}_3\text{COCH}_3][\text{HCl}]$$
$$k = \frac{\text{rate}}{[\text{CH}_3\text{COCH}_3][\text{HCl}]}$$
$$k = \frac{1.1 \times 10^{-5}\,\text{mol}\,\text{dm}^{-3}}{0.5\,\text{mol}\,\text{dm}^{-3} \times 1.25\,\text{mol}\,\text{dm}^{-3}}$$
$k = 2 \times 10^{-5}\,\text{mol}\,\text{dm}^{-3}$ (only 1 significant figure since data to only one significant figure).
(c) $\text{mol}^{-1}\,\text{dm}^3\,\text{s}^{-1}$.

Reaction rates and temperature

Checkpoints

1 (a) Draw a black cross on a piece of paper. Put the paper under the reaction vessel and look at the cross through the reaction mixture. Record the time taken for the milkiness to obscure the cross from view.
  (b)

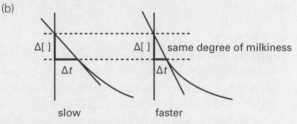

From a graph of concentration [ ] against time t, rate $\propto \Delta[\ ]/\Delta t$ but $\Delta[\ ]$ is constant, so rate $\propto 1/\Delta t$. The higher the rate, the smaller Δt.
2 If E_a increases k will decrease.
  If temperature decreases k will decrease.
3 (a) $\Delta H = E_{a(\text{forward})} - E_{a(\text{reverse})}$
    Note that this expression always gives the correct sign for ΔH because activation energy values are always positive.
  (b) For an exothermic reaction the energy of the products is lower than the energy of the reactants and so the activation energy of the reverse reaction must be greater than the activation energy of the forward reaction. Thus the forward (exothermic) reaction is kinetically more likely to occur than the reverse (endothermic) reaction.

Exam question

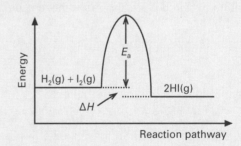

The difference between the activation energy of the forward and reverse reactions is very small, so the reaction should be reversible. The activation energies are high, so the reaction should be slow to attain equilibrium.

Catalysis

Checkpoints

1 (a) $N_2(g) + 3H_2(g) \rightleftharpoons 2NH_3(g)$; iron
  (b) $2SO_2(g) + O_2(g) \rightleftharpoons 2SO_3(g)$; vanadium(V) oxide
2 The activation energy will be high since two negatively charged particles have to approach each other and come together. The like charges will repel and tend to keep them apart.

Exam questions

1 (a) The charge density of the Fe^{2+} ion and of the peroxodisulphate ion might be lower than the charge densities of the Fe^{3+} ion and the iodide ion. If so, the Fe^{3+} ion and iodide ion would attract each other more strongly than would the Fe^{2+} ion and peroxodisulphate ion. This might make the reaction between Fe^{3+} ions and iodide ions faster than the reaction between Fe^{2+} ions and peroxodisulphate ions. However, both reactions will have low activation energies and be fast.
  (b) No effect since we have the same reactant and the same products.
  (c)

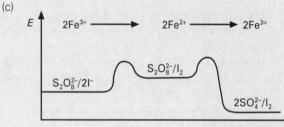

Reaction pathway

2 $5C_2O_4^{2-}(aq) + 16H^+(aq) + 2MnO_4^-(aq)$
  $\rightarrow 10CO_2(g) + 2Mn^{2+}(aq) + 8H_2O(l)$
A product, Mn^{2+} ions, catalyses the reaction.
Heat from the redox reaction raises the temperature.

Structured exam question

(a)

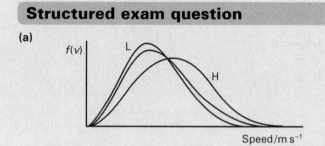

Speed/m s^{-1}

Notice that the distribution curve gets lower and fatter as the temperature rises *but* the area under the curve does *not* change because it is proportional to the total number of molecules.

(b) (i) The activation energy E_a is the minimum energy needed by reacting particles (atoms, ions or molecules) to achieve the transition state so that a reaction may occur between them.

(b) (ii), (iii)

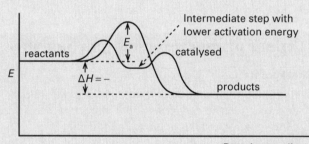

Reaction cordinate

(iv) A catalyst provides an alternative route for the reaction. This has the net effect of lowering the overall activation energy for the reaction.
Consequently a greater proportion of the reacting particles (atoms, ions or molecules) will have the minimum energy, E_a, needed for a reaction to take place.

Equilibria

You read and write equations from left to right in the direction → of the chemical change or forward reaction. You may know that some chemical changes can also go backwards ← or in the reverse direction. In this chapter you will learn about reversible reactions ⇌ which include mixtures of gases and, more importantly, acids and bases in aqueous solution. You must be able to apply the law of chemical equilibrium to these reactions. First make sure you understand the principles behind the law before you worry about the maths involved.

Exam themes

→ Describe and explain dynamic equilibria and apply Le Chatelier's principle to reactions at equilibrium

→ Write expressions for equilibrium constants K_c and K_p, and using supplied data, calculate values for K_c or work out equilibrium concentrations of reactants and products

→ Calculate partial pressures of gases in order to calculate K_p, and given a value of K_p work out partial pressures of products or reactants at equilibrium

→ Define and give examples of strong and weak acids and bases and give a Brønsted–Lowry definition of acid and base

→ Recognize conjugate acids and bases and work out pH of solutions from concentrations of $H^+_{(aq)}$ or $OH^-_{(aq)}$

→ Give expressions for K_a and K_b and use supplied values of K_a and K_b (or pK_a and pK_b) to work out the pH of solutions

→ Sketch pH curves and interpret ones supplied to choose a suitable indicator for a titration

→ Define a buffer, describe how it works and calculate its pH

Topic checklist

O AS ● A2	AQA	CCEA	EDEXCEL	OCR	WJEC
Dynamic equilibria	O	O	O	O	O
Chemical equilibria	●	●	●	●	●
Equilibrium problems	O●	O●	O●	O●	O●
Aqueous equilibria	●	●	●	●	●
Weak acids and bases	●	●	●	●	●
Acid–base reactions	●	●	●	●	●
Buffers and indicators	●	●	●	●	●

Dynamic equilibria

Reactions usually carry on until all of the reactants have turned into products or until one or other of the reactants runs out, but this isn't always the case. Take, for example, the reaction between nitrogen and hydrogen to make ammonia. As the ammonia builds up it starts to decompose back to nitrogen and hydrogen. Eventually, the reaction seems to stop with only about 10% of the elements converted into ammonia under normal conditions. The forward and reverse reactions are occurring at the same rate and a dynamic equilibrium is reached.

Phase equilibria ●●●

The melting point and boiling point of elements and compounds are two characteristic properties listed in data books.

→ The melting point of a substance is the temperature at which the solid and liquid phases can coexist in equilibrium (at 1 atm).

When a solid gains energy, its particles (atoms, ions or molecules) vibrate more vigorously about their fixed positions in the lattice. When a solid gains enough energy to melt, its particles leave the lattice and their kinetic energy changes from vibrational energy into translational and rotational energy. When a liquid (or solid) evaporates, its particles escape from the surface to form a vapour.

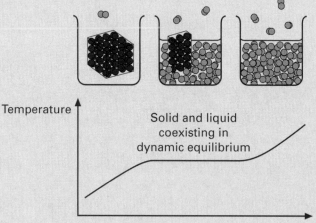

The converse is true when a vapour condenses and a liquid solidifies.

In an open vessel, liquids evaporate as their particles escape into the surrounding air. In a closed vessel at a fixed temperature, liquid and vapour attain dynamic equilibrium. The rate at which the particles escape from the liquid equals the rate at which they return from the vapour.

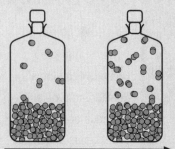

Increasing temperature

→ The vapour pressure of a liquid increases with increasing temperature.

→ The boiling point of a substance is the temperature at which the vapour pressure of the liquid equals 1 atm (101 325 Pa) or the external pressure when it is not 1 atm.

Chemical equilibria ●●●

Reversible reactions are chemical changes that

→ may take place in either direction depending upon the conditions
→ may be homogeneous (one phase) or heterogeneous (two or more phases)
→ may attain dynamic equilibrium in which the rate of the forward reaction equals the rate of the reverse reaction

You will meet a variety of reversible reactions including the following:

$$CH_3CO_2C_2H_5(l) + H_2O(l) \rightleftharpoons CH_3CO_2H(l) + C_2H_5OH(l)$$
$$N_2O_4(g) \rightleftharpoons 2NO_2(g)$$
$$N_2(g) + 3H_2(g) \rightleftharpoons 2NH_3(g); \Delta H^\ominus = -94.4 \text{ kJ mol}^{-1}$$
$$2SO_2(g) + O_2(g) \rightleftharpoons 2SO_3(g); \Delta H^\ominus = -197 \text{ kJ mol}^{-1}$$

Law of chemical equilibrium ●●●

In general, a reversible reaction can be represented by the equation

$$aA + bB \rightleftharpoons cC + dD$$

At equilibrium at a constant temperature T, then $[C]^c \times [D]^d$ divided by $[A]^a \times [B]^b$ has a constant value, K_c, if the chemicals are all aqueous or in the liquid state. If the chemicals are in the gaseous state, then

$$\frac{p_C^c \times p_D^d}{p_A^a \times p_B^b} = K_p \quad \text{where } p \text{ is the partial pressure}$$

The letter K stands for the value of the constants, usually called equilibrium constants. If we use a subscript (eq) to show that the concentration or partial pressure values are the values at equilibrium, then we can write

$$K_c = \frac{[C]_{eq}^c \times [D]_{eq}^d}{[A]_{eq}^a \times [B]_{eq}^b} \quad \text{or} \quad K_p = \frac{p_{C,eq}^c \times p_{D,eq}^d}{p_{A,eq}^a \times p_{B,eq}^b}$$

The actual value and units of the equilibrium constants, K_c and K_P, depend on the balanced equation for the reversible reaction. The value of K_c or K_p for a reaction is the reciprocal of K_c and K_p for its reverse reaction and the units alter accordingly.

Watch out!

Do not use [ ] when writing expressions for K_p.

Links

See page 60: aqueous equilibria; page 127: ligand replacement reactions of complex ions.

Checkpoint 2

What would be the units of k_c and K_p if (i) $a + b = c + d$, (ii) $a = 1$, $b = 3$, $c = 2$, $d = 0$?

Exam question (5 min)   answer: page 69

(a) What is meant by 'standard molar enthalpy change of vaporization'?

(b) Suggest why the value for water is almost seven times that for hexane.

(c) Explain why water boils below 100 °C at a high altitude in the Himalayas.

Chemical equilibria

So what do K_c and K_p tell us and how can we change them? The bigger K_c and K_p are, the higher the conversion rate of reactants to products and the more products we expect to get. Only one factor can change the size of K_c and K_p for a specified reaction and that is the temperature; in this section we show you how.

The jargon

When the value of K is very small we say that *the (position of) equilibrium lies to the left*. When the value of K is very large we say *the equilibrium lies well to the right*.

Factors affecting equilibrium constant values ●●●

Nature of the reversible reaction

The value represented by K_c and K_p for a given temperature depends upon the chemical system being considered.

Chemical system at 500 K	K_p	Units	$\Delta H^{\ominus}_{500}$/kJ mol^{-1}
$H_2(g) + CO_2(g) \rightleftharpoons H_2O(g) + CO(g)$	7.8×10^{-3}	–	$+41$
$N_2(g) + 3H_2(g) \rightleftharpoons 2NH_3(g)$	3.8×10^{-2}	atm^{-2}	-101
$H_2(g) + I_2(g) \rightleftharpoons 2HI(g)$	25.0	–	-10
$N_2O_4(g) \rightleftharpoons 2NO_2(g)$	1.7×10^3	atm	$+57$
$2SO_2(g) + O_2(g) \rightleftharpoons 2SO_3(g)$	2.5×10^{10}	atm^{-1}	-200

From the above K_p values we can tell that at 230 °C very little reaction occurs between hydrogen and carbon dioxide or nitrogen, whereas the dissociation of dinitrogen tetraoxide is considerable and the formation of sulphur trioxide is almost complete.

Watch out!

In any chemical equation we always write the products on the right-hand side (and reactants on the left-hand side). In the equilibrium constant expression we always write the [ ] or p terms for the products on the top (the numerator) and the [ ] or p terms for reactants on the bottom (the denominator).

The chemical equation

The value represented by K_c and K_p for a given temperature depends upon the way we write the equation for the reaction.

T/°C	Balanced equation	K_p	Units	K_c	Units
80	$N_2O_4(g) \rightleftharpoons 2NO_2(g)$	4	atm	0.138	mol dm^{-3}
80	$\frac{1}{2}N_2O_4(g) \rightleftharpoons NO_2(g)$	2	atm$^{1/2}$	0.371	mol$^{1/2}$ dm$^{-1\frac{1}{2}}$
80	$NO_2(g) \rightleftharpoons \frac{1}{2}N_2O_4(g)$	$\frac{1}{2}$	atm$^{-1/2}$	2.70	mol$^{-1/2}$ dm$^{1\frac{1}{2}}$
80	$2NO_2(g) \rightleftharpoons N_2O_4(g)$	$\frac{1}{4}$	atm^{-1}	7.25	mol^{-1} dm^3

Temperature of the equilibrium system

The value represented by K_c and K_p for a given chemical system depends upon the temperature at which the system is being kept.

For the dissociation $N_2O_4(g) \rightleftharpoons 2NO_2(g)$; $\Delta H^{\ominus} = +58$ kJ mol^{-1},

The jargon

Dissociation is the splitting of a molecule (or ion) into smaller molecules (or ions). It is usually an endothermic process brought about by the action of heat upon a substance (or by the action of polar solvents upon weak electrolytes).

K_p/atm:	0.115	3.89	47.9	347	1 700	6 030
T/°C:	25	77	127	177	227	277

If we warm an equilibrium mixture of dintrogen tetraoxide and nitrogen dioxide (NO_2, brown), we see the colour become darker brown as the K_p value increases. If we cool the mixture, we see the colour become lighter brown as the K_p value decreases.

→ K_c and K_p for the *endothermic* direction of a reaction will *increase* with increasing temperature.

→ K_c and K_p for the *exothermic* direction of a reaction will *decrease* with increasing temperature.

Factors affecting the composition of a system

Total pressure

→ If the total number of moles of gaseous products is greater than that of the gaseous reactants, an increase in the total pressure is accompanied by a shift in the composition in favour of reactants.

Composition of equilibrium mixture of $N_2O_4(g)$ and $NO_2(g)$ at 80 °C

p N_2O_4/atm	p NO_2/atm	Total pressure/atm	mol % NO_2	mol % N_2O_4
0.17	0.83	1	83	17
0.54	1.46	2	73	27
1.00	2.00	3	67	33
1.53	2.47	4	62	38

→ If the total number of moles of gaseous products is the same as that of the gaseous reactants, the equilibrium composition does not change when the total pressure changes.

Partial pressure (or concentration)

→ If the partial pressure (or concentration) of a component is altered, the equilibrium may be disturbed and the composition may alter to restore equilibrium.

The following data show the result of adding extra iodine to an equilibrium mixture of iodine, hydrogen and hydrogen iodide.

Composition of equilibrium mixture of $H_2(g)$, $I_2(g)$ and $HI(g)$ at 227 °C

	Amount I_2 added/mol	$I_2(g)$/mol	$H_2(g)$/mol	$HI(g)$/mol	K
(a)	0	1.000	1.000	5.000	25.00
(b)	1	1.720	0.719 5	5.561	25.00
(c)	2	2.548	0.547 5	5.905	25.00
(d)	4	4.361	0.361 4	6.277	25.00

Add 1 mol I_2 to 1 mol I_2 in a flask on its own and you have 2 mol I_2 in the flask. Add 1 mol I_2 to 1 mol I_2 in equilibrium with 1 mol H_2 and 5 mol HI and you have less than 2 mol I_2 in the flask. Why? Because some of the added iodine combines with (and decreases) the hydrogen present to produce extra hydrogen iodide.

→ If in a reversible reaction at equilibrium the partial pressure (or concentration) of one reactant is increased, the partial pressures of the other reactants decrease and the partial pressures of the products increase to restore the chemical equilibrium.

Checkpoint 1

Use the law of chemical equilibrium to explain why the industrial synthesis of ammonia from nitrogen and hydrogen by the Haber process is carried out at a total pressure of about 200 atm.

Watch out!

The value of K_p for the dissociation of $N_2O_4(g)$ at 80 °C is constant at 4 atm for all these equilibrium mixtures.

Checkpoint 2

Use the law of chemical equilibrium to explain why, in the Contact process for the industrial production of sulphuric acid, sulphur dioxide is reacted with an excess of air.

Links

See page 21: partial pressure.

Exam question (6 min)   answer: page 69

State Le Chatelier's principle. Write an expression for K_p for the endothermic dissociation of $PCl_5(g)$ into $PCl_3(g)$ and $Cl_2(g)$. Explain the effect upon the dissociation of increasing (i) the total pressure and (ii) the temperature.

Equilibrium problems

You should be able to predict and explain changes in chemical equilibria with temperature, pressure or concentration. You should also be able to solve simple problems involving K_c, K_p and equilibrium composition.

Le Chatelier's principle   ●●●

The law of chemical equilibrium deals *quantitatively* with the way the composition of an equilibrium system may change with temperature, pressure and concentration. In 1888 Henri Le Chatelier put forward his principle which dealt *qualitatively* with the effects of stress upon a system in equilibrium. There are probably as many different English translations as there are textbooks.

→ If you alter the conditions of a reversible reaction and disturb the equilibrium, the composition of the mixture may change to restore the equilibrium and to minimize the effect of altering the conditions.

According to Le Chatelier's principle, the yield of sulphur trioxide in the Contact process would be increased by increasing the total pressure because the formation of $SO_3(g)$ is accompanied by a decrease in volume (3 volumes to 2 volumes): $2SO_2(g) + O_2(g) \rightleftharpoons 2SO_3(g)$.

Problems   ●●●

Calculating a value for K_c or K_p

Example: The degree of dissociation of dinitrogen tetraoxide into nitrogen dioxide is 0.5 at 60 °C and a total pressure of 1 atm: $N_2O_4(g) \rightleftharpoons 2NO_2(g)$. Calculate the equilibrium constant K_p at this temperature.

$$N_2O_4(g) \rightarrow 2NO_2(g)$$
$$1 \text{ mol} \rightarrow 2 \text{ mol}$$

so

$$0.5 \text{ mol} \rightarrow 1 \text{ mol}$$

If 0.5 of 1 mol N_2O_4 dissociates, the resulting gas mixture will contain 0.5 mol (undissociated) N_2O_4 and 1 mol NO_2, so the amounts of the two gases will be in the ratio $1:2$.

$$\text{partial pressure } N_2O_4 = {}^1/_3 \text{ atm}; \quad NO_2 = {}^2/_3 \text{ atm}$$

$$K_p = \frac{p^2_{NO_2}}{p_{N_2O_4}} = \frac{\left({}^2/_3\right)}{\left({}^1/_3\right)}$$

Hence $K_p = 1{}^1/_3$ atm.

Calculating the composition of a mixture   ●●●

Example: At a temperature of 650 °C and a total pressure of 3 atm, dimeric aluminium chloride dissociate into monomers according to the equation $Al_2Cl_6(g) \rightleftharpoons 2AlCl_3(g)$. The value of K_p is 4.0 atm at 650 °C. Calculate the percentage dissociation of the aluminium chloride.

$$Al_2Cl_6(g) \rightarrow 2AlCl_3(g)$$
$$1 \text{ mol} \rightarrow 2 \text{ mol}$$

If x mol dissociates then the composition of the mixture will be

$$(1 - x) \text{ mol} + 2x \text{ mol} = (1 + x) \text{ mol total}$$

and the partial pressures of the components will be

$$p_d = 3 \times (1 - x)/(1 + x) \text{ and } p_m = 3 \times 2x/(1 + x)$$

K_p is

$$
\begin{aligned}
(p_m)^2/p_d = 4.0 &= [3 \times 2x/(1 + x)]^2/[3 \times (1 - x)/(1 + x)] \\
4.0 \times [3 \times (1 - x)/(1 + x)] &= [3 \times 2x/(1 + x)]^2 \\
4.0 \times 3 \times (1 - x) \times (1 + x) &= (3 \times 2x)^2 \\
12.0 \times (1 - x^2) = 36x^2 \qquad 12.0 &- 12.0x^2 = 36x^2 \\
12.0 = 48x^2 \qquad 1.0 &= 4x^2
\end{aligned}
$$

which simplifies to

$$x = \tfrac{1}{2}$$

Aluminium chloride is 50% dissociated.

Example: Water and carbon monoxide in the molar ratio of $1:1$ are reacted together at $600\,°C$. The equation for the reaction is $H_2O(g) + CO(g) \rightleftharpoons H_2(g) + CO_2(g)$ and the value of the equilibrium constant is 2.9 at $600\,°C$. Calculate the mole percentage of hydrogen in the mixture at equilibrium.

$$H_2O(g) + CO(g) \rightarrow H_2(g) + CO_2(g)$$
$$1 \text{ mol} + 1 \text{ mol} \rightarrow 1 \text{ mol} + 1 \text{ mol}$$

Let $V \text{ dm}^3$ equilibrium mixture have x mol H_2. Then the composition of the mixture will be

$$(1 - x) + (1 - x) \rightarrow x + x$$

and the concentrations of the components will be

$$(1 - x)/V + (1 - x)/V \rightarrow x/V + x/V$$

So

$$\frac{(x/V)(x/V)}{[(1 - x)/V][(1 - x)/V]} = 2.9$$

which simplifies to

$$x^2/(1 - x)^2 = 2.9$$

Take the square root of both sides

$$x/(1 - x) = +1.70 \text{ (why reject } -1.70?)$$
$$x = 0.63$$
$$H_2O(g) + CO(g) \rightarrow H_2(g) + CO_2(g)$$
$$0.37 \qquad 0.37 \qquad 0.63 \qquad 0.63 \quad = 2 \text{ mol total}$$

0.63 mol H_2 in 2 mol mixture, 31.5% by mole of hydrogen at equilibrium.

Exam question (5 min) answer: page 69

2 mol ethanoic acid $CH_3CO_2H(l)$ and 1 mol ethanol $C_2H_5OH(l)$ react to form an equilibrium mixture containing 0.85 mol ethyl ethanoate at $60\,°C$. Write an equation for the esterification reaction and calculate the value of K_c.

Watch out!

This first question is easier than the next one because you do not need to solve a quadratic equation.

Examiner's secrets

You could omit this working from your answer. These lines are included in case you would like some help with the maths.

Examiner's secrets

A-level questions don't come much harder than this but you can still score quite a few marks even if you do not solve the equation. This is the kind of working you should show.

'Which simplifies to' means the Vs cancel to give x^2 at the top (numerator) and at the bottom (denominator) the expression $(1 - x)(1 - x)$.

It is unlikely that you will be given examples which need you to solve a quadratic equation.

You should spot that you can take the square root of both sides. So $x/(1 - x) - \sqrt{(2.9)}$. This is a simple equation in x. If, say, the mole ratio $3:1$ for H_2O and CO had been used then you would have to have solved a quadratic equation.

Aqueous equilibria

The law of chemical equilibrium applies to the ionization of water and to aqueous acid–base reactions. K_w, K_a and K_b are three important constants.

Ionization of water

The fundamental equilibrium that exists in water and all aqueous solutions is $H_2O(l) + H_2O(l) \rightleftharpoons H_3O^+(aq) + OH^-(aq)$. When we apply the law of chemical equilibrium (and some simplifying mathematical assumptions) to this proton-transfer reaction we can write

$$[H_3O^+(aq)][OH^-(aq)] = K_w$$
$$K_w = 1 \times 10^{-14} \text{ (or 0.000 000 000 000 01) } mol^2\,dm^{-6} \text{ at } 25\,°C$$

Even expressed in standard form as a negative power of 10, the value of K_w is an inconvenient number. So we write

$$pK_w = -\log_{10}(K_w/mol^2\,dm^{-6}) = 14 \text{ at } 25\,°C$$

In pure water the concentration of $H_3O^+(aq)$ must equal the concentration of $OH^-(aq)$. So it follows that

$$[H_3O^+(aq)] = [OH^-(aq)] = 1 \times 10^{-7} \text{ mol\,dm}^{-3} \text{ at } 25\,°C$$

For convenience, we write

$$pH = -\log_{10}([H_3O^+(aq)]/mol\,dm^{-3})$$
$$pOH = -\log_{10}([OH^-(aq)]/mol\,dm^{-3})$$

These two definitions combine with the definition of pK_w to give the following important relationship:

➡ $pK_w = pH + pOH = 14$ at $25\,°C$

pH scale of acidity

An aqueous solution is neutral if $pH = pOH = 7$ at $25\,°C$.

Acid solutions

Hydrogen chloride gas dissolves rapidly in water and ionizes completely: $HCl(g) + H_2O(l) \rightarrow H_3O^+(aq) + Cl^-(aq)$.
    Consequently, in a $0.1\ mol\,dm^{-3}$ solution of hydrochloric acid,

$$[H_3O^+(aq)] = 1 \times 10^{-1}\ mol\,dm^{-3}$$

Therefore $pH = -\log_{10}([H_3O^+(aq)]/mol\,dm^{-3}) = 1$. But $pH + pOH = 14$ at $25\,°C$. Therefore $pOH = 13$.

➡ An aqueous solution is acidic if $pH < 7$.

Alkaline solutions

Sodium hydroxide pellets dissolve rapidly in water to give aqueous sodium ions and hydroxide ions: $NaOH(s) \rightarrow Na^+(aq) + OH^-(aq)$. Consequently, in a $0.1\ mol\,dm^{-3}$ solution of sodium hydroxide,

$$[OH^-(aq)] = 1 \times 10^{-1}\ mol\,dm^{-3}$$

Therefore $pOH = -\log_{10}([OH^-(aq)]/mol\,dm^{-3}) = 1$. But $pH + pOH = 14$ at $25°C$. Therefore $pH = 13$.

➡ An aqueous solution is alkaline if $pH > 7$.

pH range

In principle we could have acidic solutions with pH < 0 and basic solutions with pOH < 0. In practice we work mostly with solutions in the pH range from 1 to 13.

$$[H_3O^+(aq)] \rightarrow 10^{-1} \qquad\qquad 10^{-7} \qquad\qquad\qquad 10^{-13}$$

```
[H3O+(aq)] → 10⁻¹              10⁻⁷                  10⁻¹³
/mol dm⁻³     |                 |                      |
       pH   1  2  3  4  5  6  7  8  9  10 11 12 13
```

acidic	neutral	alkaline
$[H^+] > [OH^-]$	$[H^+] = [OH^-]$	$[H^+] < [OH^-]$

```
            13 12 11 10  9  8  7  6  5  4  3  2  1   pOH
             |                 |                 |
            10⁻¹³            10⁻⁷             10⁻¹ ← /mol dm⁻³
```
[OH⁻(aq)]

Strong acids and bases

Mineral acids such as hydrochloric, nitric and sulphuric acid are strong acids and the s-block metal hydroxides such as sodium, potassium, calcium and barium hydroxide are strong alkalis.

→ Strong acids and strong bases are completely ionized in water.

Calculations

You could be asked to do simple calculations involving pH and the composition of aqueous strong acids and bases. The problems fall into two types. Here is an example of each type.

Example 1:   What is the concentration of dilute nitric acid of pH = 1.4?

$$pH = 1.4 = -\log_{10}([H_3O^+(aq)]/mol\,dm^{-3})$$
$$[H_3O^+(aq)] = 4 \times 10^{-2}\,mol\,dm^{-3}$$

but nitric acid is strong: $HNO_3(aq) \rightarrow H^+(aq) + NO_3^-(aq)$. Therefore the concentration of the nitric acid is $4 \times 10^{-2}\,mol\,dm^{-3}$ or $0.04\,mol\,dm^{-3}\,HNO_3(aq)$.

Example 2:   What is the pH of $0.2\,mol\,dm^{-3}\,Ba(OH)_2(aq)$?

$$0.02\,mol\,Ba(OH)_2(aq) \rightarrow 0.02\,mol\,Ba^{2+} \text{ but } 0.04\,mol\,OH^-$$
$$[OH^-(aq)] = 4 \times 10^{-2}\,mol\,dm^{-3}$$

but

$$pOH = -\log_{10}([OH^-(aq)]/mol\,dm^{-3})$$
$$pOH = 1.4$$

but

$$pOH + pH = 14$$
$$pH = 12.6$$

Exam question (6 min)

answer: page 70

(a) Define pH.

(b) Calculate the approximate pH of aqueous sulphuric acid of concentration $0.05\,mol\,dm^{-3}$ stating clearly any assumptions made.

(c) Calculate the concentration, in $g\,dm^{-3}$, of calcium hydroxide in limewater whose pH is 10.8.

Watch out!

$Ba(OH)_2(s)$ is an ionic solid 0.1 mol of which gives 0.2 mol $OH^-(aq)$ when it is dissolved in water.
$H_2SO_4(l)$ is a covalent liquid 0.1 mol of which gives 0.1 mol $H_3O^+(aq)$ when it reacts with water:
$$H_2SO_4(aq) \rightarrow H^+(aq) + HSO_4^-(aq)$$

The jargon

We may simplify an equation like
$$HNO_3(aq) + H_2O(l) \rightarrow H_3O^+(aq) + NO_3^-(aq)$$
by writing
$$HNO_3(aq) \rightarrow H^+(aq) + NO_3^-(aq)$$

Examiner's secrets

You will be given simple numbers and easy arithmetic where possible. When you use your calculator for the logs (base 10 remember) round your answer to an appropriate significant figure or you may lose marks!

The jargon

Limewater is a saturated aqueous solution of calcium hydroxide.

Checkpoint 4

What gas is identified using limewater and how does the test work?

Weak acids and bases

The law of chemical equilibrium applies to the partial ionization of weak acids and bases. The dissociation constants, K_a and K_b, are important constants.

Dissociation of weak acids and bases ●●●

Weak acids and bases are only partially ionized in water.

Weak acids

Liquid ethanoic acid mixes in all proportions with water but fewer than 10% of its molecules ionize into hydrogen ions and ethanoate ions:

$$CH_3CO_2H(aq) + H_2O(l) \rightleftharpoons H_3O^+(aq) + CH_3CO_2^-(aq)$$

When we apply the law of chemical equilibrium (and some simplifying mathematical assumptions) to this proton-transfer reaction we can write

$$\frac{[H_3O^+(aq)][CH_3CO_2^-(aq)]}{[CH_3CO_2H(aq)]} = K_a \text{ (the acid dissociation constant)}$$

At 25 °C the value of K_a for ethanoic acid is 1.7×10^{-5} mol dm^{-3}. We can obtain a more convenient number by defining

→ $pK_a = -\log_{10}(K_a/\text{mol dm}^{-3})$
→ pK_a for ethanoic acid = 4.8 and is typical for carboxylic acids

Weak bases

Ammonia gas dissolves rapidly in water but less than 10% of the dissolved ammonia ionizes into ammonium ions and hydroxide ions:

$$NH_3(aq) + H_2O(l) \rightleftharpoons NH_4^+(aq) + OH^-(aq)$$

When we apply the law of chemical equilibrium (and some simplifying mathematical assumptions) to this proton-transfer reaction we can write

$$\frac{[NH_4^+(aq)][OH^-(aq)]}{[NH_3(aq)]} = K_b \text{ (the base dissociation constant)}$$

At 25 °C the value of K_b for aqueous ammonia is 1.8×10^{-5} mol dm^{-3}. We can obtain a more convenient number by defining

→ $pK_b = -\log_{10}(K_b/\text{mol dm}^{-3})$
→ pK_b for aqueous ammonia = 4.8 and is similar for alkyl amines

Calculations

You could be asked to do calculations involving pH, K_a, K_b and the composition of a solution. At A-level you can usually make the following simplifying assumptions.

1 The weak acid (or alkali) supplies all the H$_3$O$^+$(aq) ions (or all the OH$^-$(aq) ions) for the solution, the amount coming from the water being negligible.
2 The weak acid (or alkali) ionizes so little that it can be regarded as being un-ionized when calculating the concentration of its (undissociated) molecules.

There are three types of problem.

1. Calculate pH from the concentration and K_a (or K_b) of an aqueous weak acid (or alkali). This is very popular in examinations.
2. Calculate K_a (or K_b) of a weak acid (or alkali) from the concentration and pH of its solution. This occurs in connection with titration curves.
3. Calculate the concentration of an aqueous weak acid (or alkali) from its K_a (or K_b) and pH of the solution. This occurs with buffer solutions.

Here is an example of type 1.

Example: What is the pH of an aqueous solution of ethanoic acid of concentration $0.1\ \text{mol dm}^{-3}$? ($K_a = 1.7 \times 10^{-5}\ \text{mol dm}^{-3}$)

$$CH_3CO_2H(aq) + H_2O(l) \rightleftharpoons H_3O^+(aq) + CH_3CO_2^-(aq)$$
$$\frac{[H_3O^+(aq)][CH_3CO_2^-(aq)]}{[CH_3CO_2H(aq)]} = 1.7 \times 10^{-5}\ \text{mol dm}^{-3}$$

If the acid provides all the $H_3O^+(aq)$ ions then

$$[H_3O^+(aq)] = [CH_3CO_2^-(aq)]$$

If the acid dissociation into ions is negligible then

$$[CH_3CO_2H(aq)] = 0.1\ \text{mol dm}^{-3}$$
$$[H_3O^+(aq)]^2 = 0.1 \times 1.7 \times 10^{-5}\ \text{mol}^2\,\text{dm}^{-6}$$
$$[H_3O^+(aq)] = \sqrt{(1.7)} \times 10^{-3}\ \text{mol dm}^{-3}$$
$$pH = -\log_{10}([H_3O^+(aq)]/\text{mol dm}^{-3}) \qquad pH = 2.9$$

Brønsted–Lowry theory

→ An acid is a molecule or ion that can donate a proton.
→ A base is a molecule or ion that can accept a proton.
→ A strong acid has a weak conjugate base and a weak acid has a strong conjugate base.

Acid–base strength

Acid	=	H⁺	+	conjugate base		$K_a/\text{mol dm}^{-3}$	pK_a
$H_3O^+(aq)$				$H_2O(l)$		1.0	0
$H_2SO_3(aq)$				$HSO_3^-(aq)$		1.5×10^{-2}	1.8
$HSO_3^-(aq)$				$SO_3^{2-}(aq)$		1.0×10^{-2}	2.0
$H_3PO_4(aq)$		Increasing acid strength		$H_2PO_4^-(aq)$	Increasing base strength	7.9×10^{-3}	2.1
$HF(aq)$				$F^-(aq)$		5.6×10^{-4}	3.3
$HNO_2(aq)$				$NO_2^-(aq)$		4.7×10^{-4}	3.3
$CH_3CO_2H(aq)$				$CH_3CO_2^-(aq)$		1.7×10^{-5}	4.8
$H_2CO_3(aq)$				$HCO_3^-(aq)$		4.3×10^{-7}	6.4
$H_2PO_4^-(aq)$				$HPO_4^{2-}(aq)$		6.2×10^{-8}	7.2
$NH_4^+(aq)$				$NH_3(aq)$		5.6×10^{-10}	9.3
$C_6H_5OH(aq)$				$C_6H_5O^-(aq)$		1.3×10^{-10}	9.9
$HPO_4^{2-}(aq)$				$PO_4^{3-}(aq)$		4.4×10^{-13}	12.4
$H_2O(l)$				$OH^-(aq)$		1.0×10^{-14}	14.0

Exam question (5 min)   answer: page 70

Using the data above:

(a) What are the values of pK_b and K_b for ammonia?

(b) Explain why an aqueous solution of trisodiumphosphate, Na_3PO_4, would be extremely alkaline.

Acid–base reactions

Watch out!

The reaction of an acid with metal is *not* a proton-transfer but an electron-transfer (redox) reaction.

Don't forget!

The lower the pK_a the stronger the acid.

Checkpoint 1

Explain how and why you can use sodium carbonate to distinguish between carboxylic acids and phenols.

Checkpoint 2

Explain why lumps of calcium oxide but not glass beads coated in concentrated sulphuric acid may be used to dry ammonia gas.

The jargon

When the pK_a is less than 7 we refer to *proton donors* and their conjugate bases as acids and their salts. When the pK_a is greater than 7 we refer to *proton acceptors* and their conjugate acids as bases and their salts.

Checkpoint 3

Explain why we should not be surprised to find that an aqueous solution of ammonium ethanoate has a pH = 7.

An acid plus a base makes a salt plus water and we call this neutralization, but the salts we make are not always neutral! We find out why and then use this information to help us to choose the right indicator for a titration.

Displacement reactions

→ A strong acid with a low pK_a can transfer its protons to the conjugate base of a weaker acid with a higher pK_a.

Acids with $pK_a < 10.3$ are stronger than the hydrogencarbonate ion (acting as a proton donor) so they will react with its conjugate base (carbonate ion), e.g.

$$CH_3CO_2H(aq) + CO_3^{2-}(aq) \rightarrow CH_3CO_2^-(aq) + HCO_3^-(aq)$$

And acids with $pK_a < 6.4$ are stronger than carbonic acid so they will react with its conjugate base (hyrogencarbonate ion), e.g.

$$CH_3CO_2H(aq) + HCO_3^-(aq) \rightarrow CH_3CO_2^-(aq) + H_2CO_3(aq)$$

The displaced carbonic acid is unstable and readily decomposes into water and carbon dioxide.

→ A strong base with a low pK_b can accept protons from the conjugate acid of a weaker base with a higher pK_b.

Alkali and alkaline earth metal hydroxides will release ammonia from ammonium salts, e.g.

$$NH_4^+(aq) + OH^-(aq) \rightarrow NH_3(aq) + H_2O(l)$$

Hydrolysis of salts

A reaction of an acid with a base is called neutralization.

Strong acids and strong bases

When hydrochloric acid reacts with aqueous sodium hydroxide according to the following equation, sodium chloride is formed and the pH of the solution is 7.

$$HCl(aq) + NaOH(aq) \rightarrow NaCl(aq) + H_2O(l)$$

But $HCl(aq) + H_2O(l) \rightarrow H_3O^+(aq) + Cl^-(aq)$ since the acid is strong and $NaOH(aq) \rightarrow Na^+(aq) + OH^-(aq)$ since the base is strong. If we omit the spectator ions $Na^+(aq)$ and $Cl^-(aq)$ the equation for the neutralization of any strong acid by any strong base may be written as

$$H_3O^+(aq) + OH^-(aq) \rightarrow H_2O(l)$$

Weak acids and weak bases

→ Aqueous salts of weak acids and strong bases have a pH > 7.
→ Aqueous salts of strong acids and weak bases have a pH < 7.
→ Aqueous salts with pH ≠ 7 are said to be hydrolysed.

You should expect an aqueous solution of ammonium chloride to have a pH < 7 because the ammonium cation, $NH_4^+(aq)$, is a moderately good

proton donor but the chloride anion, $Cl^-(aq)$, is a very weak base. You should also expect an aqueous solution of sodium ethanoate to have a pH > 7 because the sodium cation is not an acid and the ethanoate anion, $CH_3CO_2^-(aq)$, is a moderately good proton acceptor.

Acid–base titrations ●●●

Titrating is measuring with a burette the volume of acid (or alkali), of known concentration, needed to react exactly with a measured (by pipette) volume of another solution (in a conical flask) containing a small amount of an indicator.

Strong base into a strong acid

$0.1\ mol\,dm^{-3}$ NaOH(aq) titrated into 20 cm³ $0.1\ mol\,dm^{-3}$ HCl(aq)

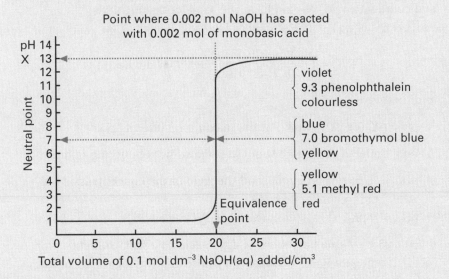

Strong base into a weak acid

$0.1\ mol\,dm^{-3}$ NaOH(aq) titrated into 20 cm³ $0.1\ mol\,dm^{-3}$ CH₃CO₂H(aq)

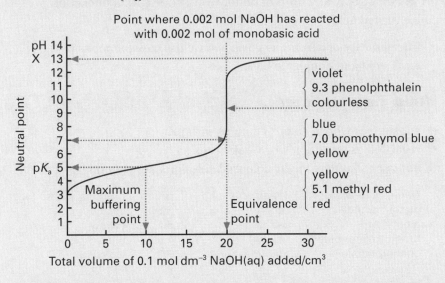

Exam question (5 min)   answer: page 70

Sketch a pH titration curve for titrating 20 cm³ $0.1\ mol\,dm^{-3}$ NH₃(aq) with $0.1\ mol\,dm^{-3}$ HCl(aq) from a burette and explain why methyl red would be a suitable indicator.

The jargon

Basicity is the number of moles of replaceable hydrogens in one mole of an acid. HCl is monobasic. H_2SO_4 is dibasic and H_3PO_4 is a tribasic acid.

Checkpoint 4

On these two diagrams what does the pH value at X correspond to?

The jargon

The *mid-point* colour of the indicator is where the pH = pK_{in}. The *end-point* is where the indicator shows its mid-point colour and the pH changes most for the smallest volume added from the burette. Ideally, the end-point and the equivalence point should be the same. Note that the end-point and equivalence point need not occur at the neutral point!

Checkpoint 5

Explain why the best indicator when titrating
(a) strong acids with strong bases is bromothymol blue
(b) weak acids by strong bases is phenolphthalein.

Buffers and indicators

A buffer is a solution whose pH is almost unchanged by the addition of small amounts of acid or alkali. Buffers contain either a weak acid and its conjugate base (e.g. ethanoic acid and ethanoate ions) or a weak base and its conjugate acid (e.g. ammonia and ammonium ions).

Buffering action

If you add acid (H^+) to, say, an ethanoic acid/sodium ethanoate buffer, the conjugate base of the buffer combines with the added hydrogen ions: $H_3O^+(aq) + CH_3CO_2^-(aq) \rightarrow CH_3CO_2H(aq) + H_2O(l)$. If you add any alkali, the acid of the buffer combines with the added hydroxide ions: $CH_3CO_2H(aq) + OH^-(aq) \rightarrow CH_3CO_2^-(aq) + H_2O(l)$. For maximum buffering effect, the weak acid and conjugate base (or weak base and conjugate acid) should have the same concentrations, i.e. $[CH_3CO_2H(aq)] = [CH_3CO_2^-(aq)]$, so they cancel. Then

$$K_a = \frac{[H_3O^+(aq)][CH_3CO_2^-(aq)]}{[CH_3CO_2H(aq)]} = [H_3O^+(aq)]$$

and therefore

→ pH = pK_a (or pOH = pK_b) at the maximum buffering point.

When a buffer is close to but not at its maximum buffering point

→ pH_{buffer} is governed by pK_a and the *ratio* of the concentrations of weak acid to conjugate base.
→ $pH_{buffer} = pK_a + \log_{10}([\text{conjugate base}]/[\text{weak acid}])$.

If you add only small amounts of dilute acid or alkali, the ratio [acid]/[conjugate base] changes very little and the logarithm of this ratio changes even less – so the buffer pH hardly changes at all. If you add large amounts of acid or alkali to a very dilute buffer solution the pH will change. A tiny drop of buffer won't cope with a bucket of concentrated acid!

→ The capacity of a buffer is governed by the *concentrations* of the weak acid and conjugate base.

Acid–base indicators

→ substances whose colour changes with pH
→ regarded as weak acids or weak bases
→ not very soluble so their coloured solutions are very dilute

Name of indicator	Weak acid form	pK_a	Conjugate base form
methyl orange	red	3.5	yellow
methyl red	red	5.1	yellow
bromothymol blue	yellow	7.0	blue
phenolphthalein	colourless	9.3	violet

Mid-point colour and pH range of an indicator

→ The dissociation constant of an indicator can be represented by

$$\frac{[H_3O^+(aq)][In^-(aq)]}{[HIn(aq)]} = K_{in}$$

The jargon

pH = $pK - \log_{10}([HA]/[A^-])$ and pH = $pK + \log_{10}([A^-]/[HA])$ are different expressions of the *Henderson–Hasselbach equation* and simply different forms of $K_a = [H^+][A^-]/[HA]$.

Checkpoint 1

Explain the effect of diluting a buffer solution by adding distilled water.

Watch out!

In titrations, we add only one or two drops of dilute indicator to the solution in our conical flask. This makes the indicator so dilute that (a) it reacts with only a negligible amount of titrant and (b) it cannot act as a buffer.

→ The colour of an acid–base indicator is governed by the ratio of the concentrations of its weak acid and conjugate base.
→ An indicator shows its mid-point colour when $[\text{HIn(aq)}] = [\text{In}^-\text{(aq)}]$ and therefore when $\text{pH} = \text{p}K_{\text{in}}$.

If HIn(aq) is yellow and In⁻(aq) is blue, when $[\text{HIn(aq)}] = [\text{In}^-\text{(aq)}]$ the mid-point colour of the indicator solution will be green.

→ An acid–base indicator changes colour over a range of about 2 pH units, one unit either side of the pH at its mid-point colour.

Selecting an indicator

For an acid–base indicator to be suitable for a titration, $\text{p}K_{\text{in}}$ shoud equal the pH at the end-point where pH changes most sharply.

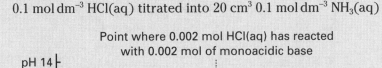
0.1 mol dm^{-3} HCl(aq) titrated into 20 cm^3 0.1 mol dm^{-3} NH₃(aq)

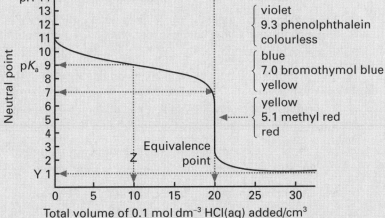

We cannot titrate a weak acid with a weak alkali, or vice versa, because the pH does not change suddenly at the end-point so no indicator will give a sharp change in colour.

We avoid using concentrations less than 0.1 mol dm^{-3} because the range over which the pH will change rapidly gets shorter with decreasing concentrations of acid and base.

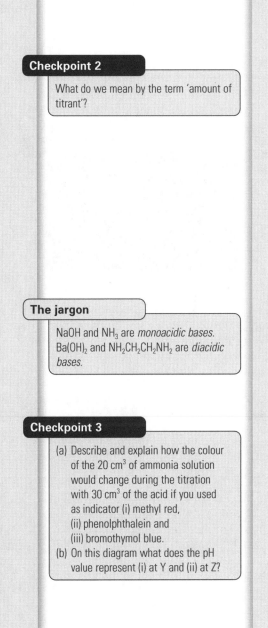

answer: page 71

Checkpoint 2

What do we mean by the term 'amount of titrant'?

The jargon

NaOH and NH₃ are *monoacidic bases*. Ba(OH)₂ and NH₂CH₂CH₂NH₂ are *diacidic bases*.

Checkpoint 3

(a) Describe and explain how the colour of the 20 cm^3 of ammonia solution would change during the titration with 30 cm^3 of the acid if you used as indicator (i) methyl red, (ii) phenolphthalein and (iii) bromothymol blue.
(b) On this diagram what does the pH value represent (i) at Y and (ii) at Z?

Exam question (7 min)

(a) (i) Write an expression for the acid dissociation constant (K_a) for the equilibrium $NH_4^+\text{(aq)} \rightleftharpoons H^+\text{(aq)} + NH_3\text{(aq)}$ and (ii) calculate $\text{p}K_a$ given that K_a is $5.6 \times 10^{-10} \text{ mol dm}^{-3}$ at $25\,°\text{C}$.

(b) Explain (i) why an aqueous solution containing both sodium dihydrogenphosphate, NaH_2PO_4, and disodium hydrogenphosphate, Na_2HPO_4, constitutes a buffer and (ii) why buffers are important in biological systems.

Structured exam question

answer: page 71

(a) Many acid–base indicators are weak acids whose aqueous solution has a different colour from that of their conjugate base. The dissociation of the indicator molecules (HInd) into their conjugate bases (Ind⁻) may be represented by

$$HInd(aq) \rightleftharpoons H^+(aq) + Ind^-(aq)$$
$$\text{colour 1} \qquad \text{colour 2}$$

The pH range of such indicators is approximately $pK_{ind} \pm 1$.

(i) Show that $pH = pK_{ind} + \log([Ind^-(aq)]/[HInd(aq)])$

....

....

....

....

....

....

(ii) If HInd(aq) is yellow, Ind⁻(aq) is blue and pK_{ind} is 4.0, what is the pH of the indicator solution whose colour is green (the mid-point colour of the indicator)?

....

(iii) If HInd(aq) is yellow, Ind⁻(aq) is blue and pK_{ind} is 7.0, state and explain what colour would be formed on adding the indicator to 0.1 mol dm⁻³ NaOH(aq).

....

....

(b) (i) State what is meant by the term *buffer solution*.

....

....

....

(ii) Explain how an aqueous solution containing ethanoic acid and sodium ethanoate behaves as a buffer solution.

....

....

....

....

....

....

(iii) Calculate the pH of a buffer solution containing 0.10 mol dm⁻³ of ethanoic acid and 0.40 mol dm⁻³ of sodium ethanoate. (The value of K_a for ethanoic acid is 1.8×10^{-5} mol dm⁻³)

....

....

....

(15 min)

Examiner's secrets

You might be given the Henderson–Hasselbalch equation in the form $pH = pK_a + \log([base]/[acid])$ or $pH = pK_a - \log([acid]/[base])$.

If not, just remember it is only an expression for K_a after taking logs of $[H^+][base]/[acid]$ and rearranging.

Examiner's secrets

If the buffer had been aqueous ammonia and ammonium chloride then $pH = pK_a - \log([NH_4^+(aq)]/[NH_3(aq)])$.

Answers
Equilibria

Dynamic equilibria

Checkpoints

1 (a) Melting point and freezing point are the same. Both terms refer to the temperature where the liquid and the solid phases are in equilibrium under the external pressure (normally atmospheric pressure).
   (b) When ice is added to a drink, some of the ice melts and this takes energy which comes from the liquid, so the temperature of the liquid falls. As long as ice remains, any energy the liquid gains from the surroundings will be used up in melting the ice so that the temperature of the drink remains low.
   (c) The molecules of the covalent compound naphthalene are held in the solid by weak van der Waals forces so the vapour pressure is high. The ions in sodium chloride are held in the lattice by strong electrostatic forces and so the vapour pressure is negligible.

2 (i) There would be no units for K_p or K_c.
   (ii) K_c would have units $mol^{-2} dm^6$ and K_p would have units pressure $^{-2}$.

Exam question

(a) The standard molar enthalpy change of vaporization is the heat change at constant pressure when one mole of liquid becomes one mole of vapour under standard thermodynamic conditions of 1 atm and 298 K.

(b) The polar water molecules are held in the liquid by van der Waals forces, permanent dipole–dipole attractions and hydrogen bonding but the non-polar hexane molecules are held in the liquid by van der Waals forces only. Therefore much more energy is needed to separate the water molecules and the boiling point is much higher.

(c) At a high altitude, the atmospheric pressure is much less than the normal 10.3 kPa at sea level. Therefore the vapour pressure of water reaches the value of the external pressure below 100 °C and the liquid boils.

Chemical equilibria

Checkpoints

1 High pressure favours a high equilibrium yield of ammonia since there are four moles of gas on the left-hand side of the equation and only two moles of gas on the right-hand side: $N_2(g) + 3H_2(g) \rightleftharpoons 2NH_3(g)$.

2 In the Contact process $2SO_2(g) + O_2(g) \rightleftharpoons 2SO_3(g)$, and at equilibrium $K_c = [SO_3]^2/[SO_2]^2[O_2]$. So more oxygen (excess) will mean less sulphur dioxide and more sulphur trioxide, i.e. more SO_2 converted into SO_3 and therefore a greater yield (equilibrium position is shifted to the right).

Exam question

Le Chatelier's principle states that if a constraint (such as change in temperature, pressure or concentration) is applied to a system in equilibrium, then the equilibrium alters in such a way as to minimize the applied constraint.

$$K_p = \frac{(p_{PCl_3} \times p_{Cl_2})}{p_{PCl_5}}$$

(i) If the total pressure is increased, the equilibrium moves to the side of the equation with the fewer number of moles of gas. In this case, this means less dissociation and less PCl_3 and Cl_2 in the equilibrium mixture.

(ii) Since the reaction in the forward direction is a dissociation it must endothermic, ΔH is positive. Therefore, a rise in temperature will favour the endothermic process and more dissociation will take place.

Equilibrium problems

Checkpoint

Although high pressure favours the formation of two moles of $SO_3(g)$ from two moles of $SO_3(g)$ and one mole of $O_2(g)$, the equilibrium yield of sulphur trioxide is very high (about 95%) even at the minimum pressure needed to drive the gases through the plant. The small increase in yield would not justify the cost of high-pressure plant.

Exam question

$CH_3COOH(l) + C_2H_5OH(l) \rightleftharpoons CH_3COOC_2H_5(l) + H_2O(l)$

initial	2 mol	1 mol		
equil.	1.15 mol	0.15 mol	0.85 mol	0.85 mol

Therefore K_c is $\dfrac{[(0.85/V) \times (0.85/V)]}{[(1.15/V) \times (0.15/V)]} = 4.2$

Aqueous equilibria

Checkpoints

1 Since the ionization of water is endothermic, an increase in temperature will increase the extent of ionization.

2 pH is $-\log_{10}(2 \times 10^{-2}) = 1.7$

3 $pOH = -\log_{10}(0.02) = 1.7$
   $pH = pK_w - pOH$
   $pH = 14 - 1.7 = 12.3$

4 The gas is carbon dioxide.
   Limewater is saturated aqueous calcium hydroxide. The acidic gas reacts with hydroxide ions to form aqueous carbonate ions that combine with the aqueous calcium ions to give the milky white precipitate of insoluble calcium carbonate:
   $CO_2(g) + 2OH^-(aq) + Ca^{2+}(aq) \rightarrow CaCO_3(s) + H_2O(l)$

Exam question

(a) $pH = -\log_{10}([H^+(aq)]/mol\,dm^{-3})$
or $pH = -\log_{10}([H_3O^+(aq)]/mol\,dm^{-3})$

(b) You could give at least two answers. You could explain that sulphuric acid is a strong monobasic acid, so
$$H_2SO_4(aq) \rightarrow H^+(aq) + HSO_4^-(aq)$$
Then $[H^+(aq)] = 0.05\,mol\,dm^{-3}$ and pH is
$-\log_{10}([H^+(aq)]/mol\,dm^{-3}) = 1.3$.

Or you could explain that sulphuric acid is a dibasic acid,
$$H_2SO_4(aq) \rightarrow 2H^+(aq) + SO_4^{2-}(aq)$$
Then $[H^+(aq)] = 2 \times 0.05\,mol\,dm^{-3}$ and pH is
$-\log_{10}([H^+(aq)]/mol\,dm^{-3}) = 1.0$.

You would get full marks for either answer.

If you also explained that there is not enough information to solve this problem accurately because the second ionization $HSO_4^-(aq) \rightleftharpoons H^+(aq) + SO_4^{2-}(aq)$ would occur to some extent, you would get a bonus mark.

(c) If pH = 10.8 then pOH is 14 − 10.8 = 3.2. So,
$$-\log_{10}([OH^-(aq)]/mol\,dm^{-3} = 3.2$$
$$[OH^-(aq)] = 6.3 \times 10^{-4}\,mol\,dm^{-3}$$
One mole of $Ca(OH)_2$ produces two moles of OH^-.
So the concentration of $Ca(OH)_2 = 3.15 \times 10^{-4}\,mol\,dm^{-3}$.
$$M_r(Ca(OH)_2) \text{ is } 40 + (2 \times 17) = 74$$
So the concentration of $Ca(OH)_2$ is
$3.15 \times 10^{-4} \times 74 = 2.3 \times 10^{-2}\,g\,dm^{-3}$.

Weak acids and bases

Exam question

(a) pK_a for ammonia is 9.3
Therefore pK_b is 14 − 9.3 = 4.7
pK_b is $-\log K_b = 4.7$
So, K_b is $2.0 \times 10^{-5}\,mol\,dm^{-3}$ (*not* 1.99×10^{-5}).

(b) When the ionic compound $(Na^+)_3PO_4^{3-}$ dissolves in water, it releases PO_4^{3-}, a very strong base which disturbs the $H_2O(l) \rightleftharpoons H^+(aq) + OH^-(aq)$ equilibrium by removing $H^+(aq)$ to form HPO_4^{2-}. $[H^+(aq)] < [OH^-(aq)]$, pH is greater than 7 and the solution is alkaline.

Acid–base reactions

Checkpoints

1 Aqueous carboxylic acids are stronger acids than aqueous carbon dioxide; therefore when a carboxylic acid is added to aqueous sodium carbonate, carbon dioxide is liberated.

Phenols in aqueous solution are weaker acids than aqueous carbon dioxide and so no reaction takes place.

2 Ammonia is a basic gas and would not react with another base like calcium oxide (which removes moisture from the ammonia by forming calcium hydroxide) but it would react with sulphuric acid.
$$NH_3 + H_2SO_4 \rightarrow NH_4HSO_4$$

3 pK_a for ethanoic acid is about 5 and pK_b for aqueous ammonia is about 5. So although the acid and base are both weak, they are weak to a similar extent. Therefore an aqueous solution of their salt is likely to be nearly neutral, pH = 7.

4 The pH value of the aqueous sodium hydroxide.

5 (a) Bromothymol blue will change colour as soon as the pH changes to a value greater than 7, i.e. it will change colour at the equivalence point.

(b) Phenolphthalein changes colour in the pH range covered by the steep portion of the curve. This means that one drop of alkali from the burette will cause the pH value to change by several pH units and the colour of the indicator to change sharply from colourless to pink.

Exam question

pH range curve:

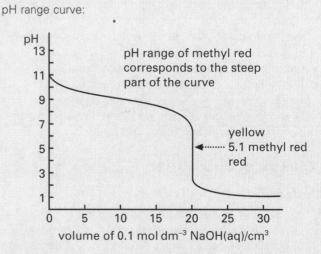

pH range of methyl red corresponds to the steep part of the curve

yellow
5.1 methyl red
red

volume of 0.1 mol dm⁻³ NaOH(aq)/cm³

Buffers and indicators

Checkpoints

1 No effect since the ratio [salt]/[acid] is constant.

2 The volume of liquid added from the burette.

3 (a) (i) Initially the ammonia solution containing methyl red would be yellow and remain yellow until near the equivalence point when it would become orange, until one drop of acid turned the solution red. No further change in colour thereafter.

(ii) The ammonia solution would remain red until about 18 cm³ of acid had been added when it

70

would change to colourless. After that no further colour change.

(iii) The ammonia solution containing bromothymol blue would start blue and when almost 20 cm^3 of acid had been added it would turn yellow and then stay yellow.

(b) Y is the pH value of the aqueous hydrochloric acid. Z is the volume of acid which gives a pH value equal to pK_a (= 9) for the *ammonium ion*. (NB pK_b is 14 − 9 = 5 and K_b = 1.0 × 10^{-5} mol dm^{-3} for aqueous ammonia.)

Exam question

(a) (i) $K_a = \dfrac{[H^+][NH_3]}{[NH_4^+]}$

(ii) pK_a = −log$_{10}$(5.6 × 10^{-10}) = 9.25

(b) (i) The hydrogenphosphate ion HPO$_4^{2-}$(aq) is the conjugate base of the dihydrogenphosphate ion H$_2$PO$_4^-$(aq). So a mixture of the two salts will work as a buffer just as ethanoic acid and its conjugate base, the ethanoate ion, act as a buffer. HPO$_4^{2-}$(aq) will accept protons from any acid added and H$_2$PO$_4^-$(aq) will donate protons to any base added to the buffer.

(ii) Buffers are important in biological systems since many biochemical processes are pH sensitive. Some enzymes only work within a narrow pH range. The blood is buffered within the range pH 7.39 – 7.41 by carbon dioxide, water and hydrogencarbonate ions.

Structured exam question

(a) (i) Applying the law of chemical equilibrium to the reversible reaction HInd(aq) ⇌ H$^+$(aq) + Ind$^-$(aq) gives

$$\frac{[H^+(aq)][Ind^-(aq)]}{[HInd(aq)]} = K_{ind}$$

Taking logarithms:

log[H$^+$(aq)] + log([Ind$^-$(aq)]/[HInd(aq)]) = log K_{ind}

Hence:

−pH + log([Ind$^-$(aq)]/[HInd(aq)]) = −pK_{ind}

Rearranging:

−pH = −pK_{ind} − log([Ind$^-$(aq)]/[HInd(aq)])

Hence:

pH = pK_{ind} + log([Ind$^-$(aq)]/[HInd(aq)])

(ii) Equal intensities of yellow and blue give green mid-point colour which means that [HInd(aq)] = [Ind$^-$(aq)] when pH is 4.0. So log([Ind$^-$(aq)])/[HInd(aq)] is log 1 (= 0) and pH = pK_{ind} = 4.0.

(iii) Blue. In 0.1 mol dm^{-3} NaOH(aq) the pH = 13 (because pOH = 1 and pH is 14 − pOH), and so

13 = 7 + log([Ind$^-$(aq)]/[HInd(aq)])

If log([Ind$^-$(aq)]/[HInd(aq)]) = 6, then

[Ind$^-$(aq)] = 10^6 × [HInd(aq)].

Blue outweighs yellow by 1 million to 1.

(b) (i) A solution of a weak acid and its conjugate base (or weak base and its conjugate acid) whose pH is almost unchanged by the addition of small amounts of acid or alkali.

(ii) In the buffer solution

CH$_3$COOH(aq) ⇌ CH$_3$COO$^-$(aq) + H$^+$(aq)

so ethanoic acid molecules are in equilibrium with ethanoate ions. When acid is added, momentarily increasing the [H$^+$(aq)], the conjugate base reacts with the added H$^+$(aq) and the equilibrium shifts to the left.

When alkali is added, momentarily increasing the [OH$^-$(aq)], the acid reacts with the added OH$^-$(aq) and the equilibrium shifts to the right.

(iii) pH = pK_a + log([CH$_3$COO$^-$(aq)]/[CH$_3$COOH(aq)])
= −log(1.8 × 10^{-5}) + log(0.40/0.10)

Thus pH = 5.3.

Electrochemistry and redox

You are already familiar with oxidation numbers in chemical names like copper(II) sulphate. In this chapter you will learn the simple rules for assigning these numbers. You will also understand the fundamentals of electrochemical cells and the importance of electrode potentials. In the previous section you saw acid–base reactions as the transfer of protons from acids to a bases. In this section you see redox reactions as the transfer of electrons from reducing agents to oxidizing agents.

Exam themes

→ Work out oxidation numbers and use them to identify redox reactions and balance redox equations

→ Describe how to measure the EMF of a cell and describe the hydrogen electrode; define what is meant by standard thermodynamic conditions; draw and interpret cell diagrams

→ Use supplied values of standard electrode and redox potentials to work out the EMF of a cell

→ Work out whether you would expect a reaction to occur using standard electrode or redox potentials

Topic checklist

O AS ● A2	AQA	CCEA	EDEXCEL	OCR	WJEC
Oxidation and reduction	O	●	O	O	O ●
Electrochemistry	●	●	●	●	●
Electrochemical series	●	●	●	●	●

Oxidation and reduction

Oxidation can be spotted in three different ways: gain of oxygen, loss of electrons or an increase in oxidation number. Not surprisingly, reduction is the reverse. We cannot have reduction without oxidation and we call reactions involving reduction and oxidation redox reactions.

Oxidation numbers ●●●

The oxidation number is a number assigned to an element according to this set of rules applied in the following priority order:

1 oxidation number of an uncombined element   0
2 sum of oxidation numbers of elements in uncharged formula   0
3 sum of oxidation numbers of elements in charged formula   charge
4 oxidation number of fluorine in any formula -1
5 oxidation number of an alkali metal in any formula $+1$
6 oxidation number of an alkaline earth metal in any formula $+2$
7 oxidation number of oxygen (except in peroxides $= -1$) -2
8 oxidation number of halogen in metal halides -1
9 oxidation number of hydrogen (except in metal hydrides $= -1$) $+1$

Redox reactions ●●●

→ Reduction is the gain of electrons by an oxidant: e.g.

$$Cl_2(aq) + 2e^- \rightarrow 2Cl^-(aq)$$

→ Oxidation is the loss of electrons by a reductant, e.g.

$$Fe^{2+}(aq) \rightarrow Fe^{3+}(aq) + e^-$$

→ Redox is the transfer of electrons from a reductant to an oxidant:

$$2Fe^{2+}(aq) + Cl_2(aq) \rightarrow 2Fe^{3+}(aq) + 2Cl^-(aq)$$

A redox reaction involves an increase ($\uparrow$) in oxidation number of one element (in the reductant) and a simultaneous balancing decrease ($\downarrow$) in oxidation number of an element (in the oxidant).

→ Disproportionation is a redox involving a simultaneous increase and decrease in oxidation number of the same element:

$$Cl_2(aq) + 2OH^-(aq) \rightarrow ClO^-(aq) + Cl^-(aq) + H_2O(l)$$

This disproportionation equation may also be written in the form

$$Cl_2(aq) + H_2O(l) \rightarrow ClO^-(aq) + Cl^-(aq) + 2H^+(aq)$$

Writing $H_2O(l)$ on the LHS and $2H^+(aq)$ on the RHS has the same effect as writing $2OH^-(aq)$ on the LHS and $H_2O(l)$ on the RHS because hydrogen ions react with hydroxide ions to form water molecules and the reaction $H^+(aq) + OH^-(aq) \rightarrow H_2O(l)$ is acid–base (*not* redox).

Balancing redox reactions ●●●

You can use the changes in oxidation numbers to work out balanced equations for redox reactions.

The jargon

In the Stock nomenclature, the *positive oxidation number* of an element is written as a Roman numeral in brackets: e.g. $CuCrO_4$ is called copper(II) chromate(VI).

Don't forget!

In Stock nomenclature, the name of a transition metal element is unchanged when the element is part of the cation but it is changed (and ends in -ate) when the element is part of the anion.

Checkpoint 1

(a) Give Stock names for
(i) Fe^{2+}, (ii) $FeCl_3$ and (iii) ClO^-.
(b) Give the oxidation number of
(i) N in HNO_3 and (ii) Mn in $KMnO_4$.
(c) Write a balanced equation for the reaction of sodium metal upon water and explain why this is a redox.

The jargon

The words *oxidant* and *reductant* can be used instead of the terms oxidizing agent and reducing agent.

Example: Reaction of acidified dichromate(VI) with sulphate(IV).

1   Write down correct formulae of reactants and products:
$$Cr_2O_7^{2-}(aq) + H^+(aq) + SO_3^{2-}(aq) \rightarrow 2Cr^{3+}(aq) + SO_4^{2-}(aq)$$

2   Apply priority rules to assign oxidation numbers:
$$Cr_2O_7^{2-}(aq) + H^+(aq) + SO_3^{2-}(aq) \rightarrow 2Cr^{3+}(aq) + SO_4^{2-}(aq)$$
$$+6 \qquad\qquad +4 \qquad +3 \qquad +6$$

3   Calculate rise and fall of oxidation numbers:

fall of $2 \times 3 = 6$
$$Cr_2O_7^{2-}(aq) + H^+(aq) + SO_3^{2-}(aq) \rightarrow 2Cr^{3+}(aq) + SO_4^{2-}(aq)$$
rise of 2

4   Balance the rise and fall of oxidation numbers:
$$Cr_2O_7^{2-}(aq) + H^+(aq) + 3SO_3^{2-}(aq) \rightarrow 2Cr^{3+}(aq) + 3SO_4^{2-}(aq)$$
$$3 \times 2 = 6$$

5   Balance other atoms *without changing balance of redox atoms*:
$$Cr_2O_7^{2-}(aq) + H^+(aq) + 3SO_3^{2-}(aq) \rightarrow 2Cr^{3+}(aq) + 3SO_4^{2-}(aq)$$

    7 oxygens   9 oxygens   12 oxygens

    (i) Write $4H_2O(l)$ on the RHS to provide 4 more oxygens.
    (ii) Write $8\,H^+(aq)$ on the LHS to balance the hydrogens.

6   Check the charges balance (same net charge on LHS and RHS):

LHS$(2-) + 8 \times (1+) + 3 \times (2-) = 0$
$$Cr_2O_7^{2-}(aq) + 8H^+(aq) + 3SO_3^{2-}(aq)$$
$$\rightarrow 2Cr^{3+}(aq) + 3SO_4^{2-}(aq) + 4H_2O(l)$$
RHS $2 \times (3+) + 3 \times (2-) = 0$

Redox titrations and calculations ●●●

The manganate(VII) ion in *excess* dilute sulphuric acid is *quantitatively* reduced to the manganese(II) ion by $Fe^{2+}(aq)$, $H_2O_2(aq)$, $C_2O_4^{2-}(aq)$, $SO_3^{2-}(aq)$, etc., and the solution changes from purple to colourless.

Consequently, we may titrate $MnO_4^-(aq)$ of known concentration from a burette into a *measured volume* of these reducing agents *without adding an indicator* to determine their concentrations.

Example: 20.0 cm³ of aqueous iron(II) sulphate need 18.5 cm³ of 0.105 mol dm⁻³ $KMnO_4(aq)$. What is the concentration of the iron(II) sulphate?

$$MnO_4^-(aq) + 8H^+(aq) + 5Fe^{2+}(aq) \rightarrow Mn^{2+}(aq) + 4H_2O(l) + 5Fe^{3+}(aq)$$
1 mol   5 mol
$18.5 \times 0.1/1000$ $5 \times (18.5 \times 0.105/1000)$ mol in 20.0 cm³ solution

So in 1000 cm³ solution $5 \times (18.5 \times 0.105/20.0) = 0.486$ mol $Fe^{2+}(aq)$.
1 mol iron(II) sulphate, $FeSO_4$, contains 1 mol iron(II) ions, Fe^{2+}.
Concentration of the iron(II) sulphate is 0.486 mol dm⁻³.

Exam question (3 min)   answer: page 81

Explain the type of reaction represented by the following equations:

(a) $Mg(s) + 2HCl(aq) \rightarrow MgCl_2(aq) + H_2(g)$
(b) $2K_2CrO_4(aq) + H_2SO_4(aq) \rightarrow K_2Cr_2O_7(aq) + K_2SO_4(aq) + H_2O(l)$

Electrochemistry

We can write separate ion–electron half-reaction equations for reduction and oxidation. We can also combine these half-equations to obtain an ionic equation for the redox.

Metal displacement reactions ●●●

Coating a nail by dipping it into aqueous copper(II) sulphate is a simple example of an electrochemical (redox) reaction.

$$Cu^{2+}(aq) + 2e^- \rightarrow Cu(s) \qquad \text{reduction}$$
$$Fe(s) \rightarrow Fe^{2+}(aq) + 2e^- \qquad \text{oxidation}$$
$$Fe(s) + Cu^{2+}(aq) \rightarrow Fe^{2+}(aq) + Cu(s) \qquad \text{redox}$$

Coating a piece of zinc by dipping it into aqueous iron(II) sulphate is another example of a metal displacement (redox) reaction.

$$Fe^{2+}(aq) + 2e^- \rightarrow Fe(s) \qquad \text{reduction}$$
$$Zn(s) \rightarrow Zn^{2+}(aq) + 2e^- \qquad \text{oxidation}$$
$$Zn(s) + Fe^{2+}(aq) \rightarrow Zn^{2+}(aq) + Fe(s) \qquad \text{redox}$$

We can use these reactions to produce an electric current.

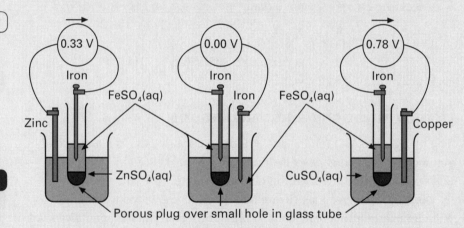

We can also use the displacement reaction of zinc with copper(II) ions.

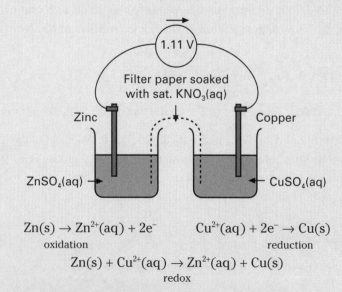

$$Zn(s) \rightarrow Zn^{2+}(aq) + 2e^- \qquad \qquad Cu^{2+}(aq) + 2e^- \rightarrow Cu(s)$$
$$\text{oxidation} \qquad \qquad \qquad \qquad \text{reduction}$$
$$Zn(s) + Cu^{2+}(aq) \rightarrow Zn^{2+}(aq) + Cu(s)$$
$$\text{redox}$$

Zinc and iron (but not copper) will also react with hydrogen ions to form hydrogen gas, e.g. $Zn(s) + 2H^+(aq) \rightarrow Zn^{2+}(aq) + H_2(g)$.

Reactivity series of metals

On the basis of their displacement reactions we can arrange the metals (and hydrogen) in reactivity order and their cations in stability order.

from most reactive $Zn(s)$ $Fe(s)$ $H_2(g)$ $Cu(s)$ to least reactive
from most stable $Zn^{2+}(aq)$ $Fe^{2+}(aq)$ $H_3O^+(aq)$ $Cu^{2+}(aq)$ to least stable

Electromotive force of electrochemical cells ⬤⬤⬤

→ The e.m.f. (E) of an electrochemical cell is the maximum potential difference (voltage) between the electrodes.
→ The *standard* e.m.f. ($E^{\ominus}$) is the maximum voltage of a cell under standard conditions, i.e. temperature 25 °C (298 K), pressure 1 atm (101 kPa) and solutions of concentration 1 mol dm^{-3}.

Apparatus diagrams and cell diagrams

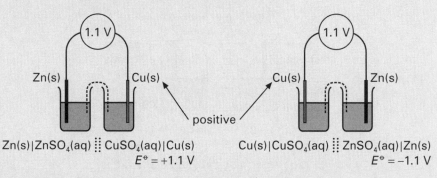

$Zn(s)|ZnSO_4(aq) \vdots\vdots CuSO_4(aq)|Cu(s)$ $Cu(s)|CuSO_4(aq) \vdots\vdots ZnSO_4(aq)|Zn(s)$
$E^{\ominus} = +1.1$ V $E^{\ominus} = -1.1$ V

The sign given to the e.m.f. associated with the cell diagram of an electrochemical cell refers to the right-hand half-cell in the cell diagram.

Standard hydrogen electrode

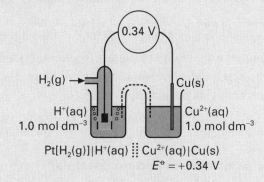

$Pt[H_2(g)]|H^+(aq) \vdots\vdots Cu^{2+}(aq)|Cu(s)$
$E^{\ominus} = +0.34$ V

The standard hydrogen electrode is the half-cell consisting of pure hydrogen gas, at 25 °C and 1 atm pressure, bubbling past a platinized platinum electrode dipping into 1.0 mol dm^{-3} $H_3O^+(aq)$.

The above value of +0.34 V is called the standard electrode potential of the system $Cu^{2+}(aq)|Cu(s)$.

Exam question (2 min)   answer: page 81

(a) For the cell shown above write the ion–electron half-equation for
    (i) the oxidation of the hydrogen gas and (ii) the reduction of the
    aqueous copper (II) ions.

(b) Combine the half-equations in (a)(i) and (ii) to write the ionic equation
    for cell redox.

Watch out!

Do not confuse the diagram of a cell (drawing of the apparatus) with a cell diagram (symbolic representation of the electrochemistry).

The jargon

In a cell diagram, a vertical line (|) represents a (phase) boundary between a solid and a solution. A pair of vertical broken lines represents a salt bridge electrolytic boundary between two solutions.

The jargon

Platinized platinum: finely divided platinum is deposited electrolytically as a black coating on the platinum surface to improve contact between the metal, the gas and the solution and to catalyse the oxidation half-reaction.

Electrochemical series

We can combine two half-cells to form a complete electrochemical cell. If one half-cell is a hydrogen electrode then the e.m.f. of the complete cell is called the electrode potential of the other half-cell.

Non-metal displacement reactions ●●●

Adding chlorine to aqueous potassium bromide or iodide produces bromine or iodine and changes the colour of the solution.

$$Cl_2(aq) + 2e^- \rightarrow 2Cl^-(aq) \qquad \text{reduction}$$
$$2Br^-(aq) \rightarrow Br_2(aq) + 2e^- \qquad \text{oxidation}$$
$$2Br^-(aq) + Cl_2(aq) \rightarrow Br_2(aq) + 2Cl^-(aq) \qquad \text{redox}$$

We can use these dispslacement reactions to produce an electric current.

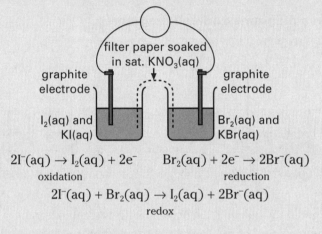

$$2I^-(aq) \rightarrow I_2(aq) + 2e^- \qquad Br_2(aq) + 2e^- \rightarrow 2Br^-(aq)$$
oxidation $\qquad\qquad\qquad$ reduction
$$2I^-(aq) + Br_2(aq) \rightarrow I_2(aq) + 2Br^-(aq)$$
redox

Reactivity series of non-metals

On the basis of their displacement reactions we can arrange the halogens in reactivity order and their anions in stability order.

| from most reactive | $F_2(g)$ | $Cl_2(aq)$ | $Br_2(aq)$ | $I_2(aq)$ | to least reactive |
| from most stable | $F^-(aq)$ | $Cl^-(aq)$ | $Br^-(aq)$ | $I^-(aq)$ | to least stable |

If we want to judge the reactivity of metals (reductants) and non-metals (oxidants) on the same basis, we should rewrite the reactivity table for non-metal anions acting as reductants

| from most reactive | $I^-(aq)$ | $Br^-(aq)$ | $Cl^-(aq)$ | $F^-(aq)$ | to least reactive |
| from most stable | $I_2(aq)$ | $Br_2(aq)$ | $Cl_2(aq)$ | $F_2(aq)$ | to least stable |

Reactivity series of metals and non-metals

If we judge the reactivity of metals and aqueous non-metal anions as reductants losing electrons, we may combine the two series into one.

from most reactive reductant					to least reactive reductant		
Zn(s)	Fe(s)	$H_2(g)$	Cu(s)	$I^-(aq)$	$Br^-(aq)$	$Cl^-(aq)$	$F^-(aq)$
$Zn^{2+}(aq)$	$Fe^{2+}(aq)$	$H_3O^+(aq)$	$Cu^{2+}(aq)$	$I_2(aq)$	$Br_2(aq)$	$Cl_2(aq)$	$F_2(aq)$
from least reactive oxidant					to most reactive oxidant		

From this series we predict that zinc and fluorine would react best to form the most stable product, zinc fluoride. In general we predict that any reductant on the left could react with any oxidant to its right.

Standard electrode potentials ●●●

The **standard electrode potential**, $E^\ominus$, is defined as the e.m.f. of an electrochemical cell in which a standard hydrogen electrode is shown as the left-hand half-cell.

The standard electrode potential for the standard hydrogen electrode is zero by definition. A typical cell diagram for the standard electrode potential of a system is $Pt[H_2(g)]|2H^+(aq) \vdots Zn^{2+}(aq) Zn(s)$; $E^{\ominus} = -0.76$ V. Data books usually give the standard reduction potential (in volts) alongside the electrode system for the right-hand half-cell.

Electrode system	$E^{\ominus}/V$
$Zn^{2+}(aq)|Zn(s)$	-0.76
$Fe^{2+}(aq)|Fe(s)$	-0.44
$2H^+(aq)|[H_2(g)]Pt$	0
$Cu^{2+}(aq)|Cu(s)$	$+0.34$
$I_2(aq), 2I^-(aq)|Pt$	$+0.54$
$Br_2(aq), 2Br^-(aq)|Pt$	$+1.09$
$Cl_2(aq), 2Cl^-(aq)|Pt$	$+1.36$

(left axis: Oxidizing, downward arrow; right axis: Reducing, upward arrow)

Combining electrode potentials

You should be able to use tables of standard electrode potentials to predict feasible reactions and to write down their cell diagram and its standard e.m.f. Here is a simple example using the table above.

RHS: reductant $Zn(s)$ is above (more powerful than) $2I^-(aq)$
LHS: oxidant $I_2(aq)$ is below (more powerful than) $Zn^{2+}(aq)$

Therefore predict $Zn(s) + I_2(aq) \rightarrow Zn^{2+}(aq) + 2I^-(aq)$.

cell diagram $Zn(s)|Zn^{2+}(aq) \vdots I_2(aq), 2I^-(aq)|Pt$
e.m.f. of cell is $(+0.54 \text{ V}) - (-0.76 \text{ V}) = +1.30$ V

Spontaneous redox reactions

→ If the e.m.f. is not zero then the electrochemical reaction will take place in the cell when its terminals are wired together.
→ Electrons will travel in the wire from the negative half-cell losing them to the positive half-cell gaining them.

Example

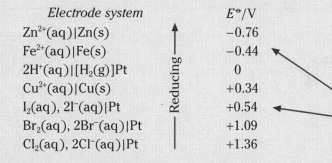

$$Zn(s)|Zn^{2+}(aq) \vdots Pb^{2+}(aq)|Pb(s) \quad E^{\ominus}_{cell} = +0.63 \text{ V}$$

Spontaneous cell reaction: $Zn(s) + Pb^{2+}(aq) \rightarrow Zn^{2+}(aq) + Pb(s)$

→ If a cell diagram corresponds to a positive e.m.f. the spontaneous cell reaction can be read from left to right.

$$Pb(s)|Pb^{2+}(aq) \vdots Zn^{2+}(aq)|Zn(s) \quad E^{\ominus}_{cell} = -0.63 \text{ V}$$

If a cell diagram corresponds to a negative e.m.f. the spontaneous cell reaction can be read from right to left.

Exam question (10 min)   answer: page 81

Use the reactivity table to predict which of the following pairs could react: (a) $H_2(g)$ and $Br_2(aq)$, (b) $Fe(s)$ and $I^-(aq)$, (c) $H_3O^+(aq)$ and $Cl_2(aq)$, (d) $Cl_2(aq)$ and $Cu^{2+}(aq)$, (e) $Cu(s)$ and $Br_2(aq)$. Explain your reasoning.

The jargon

Standard temperature for electrochemical cells is 25 °C (298 K).

Examiner's secrets

When you combine two electrode systems to form a complete cell diagram, follow these simple steps:
1 *Reverse* the *upper* system and then write it on the LHS.
2 Write the *lower* system *unchanged* on the RHS.
3 Write the symbol for the salt bridge between the two systems.
4 Check that the oxidation numbers of the molecules or ions nearest to the salt bridge symbol are higher than the oxidation numbers of the ions or molecules nearest to the terminals.

Watch out!

$Zn^{2+}(aq)|Zn(s) \vdots I_2(aq), 2I^-(aq)|Pt$ would score no marks! See secret 4 above. Make sure electrons gained in the reduction half-reaction equal the electrons lost in the oxidation half-reaction before you combine the two ion–electron half equations to get the equation for the cell reaction.

Structured exam question

answer: pages 81–2

(a) Define the term *standard electrode potential* as applied to a metallic element.

....

....

(b) The standard electrode potentials $E^{\ominus}$ of zinc and copper are $-0.76\,\text{V}$ and $+0.34\,\text{V}$ respectively.

(i) Draw an apparatus diagram to show a cell in which zinc is in contact with aqueous zinc sulphate in a beaker and copper in contact with aqueous copper(II) sulphate in a similar beaker.

(ii) State the e.m.f. of this cell?...volts.

(iii) Name the instrument that would be used to measure this e.m.f.

....

(iv) In which direction would electrons flow in a wire used to connect together the copper and zinc electrodes?

....

(v) Write an equation for the overall spontaneous cell reaction that would occur when the copper and zinc electrodes are connected together by a wire.

....

(c) The table below shows standard redox potentials of selected reductions.

Ion–electron half-equation	$E^{\ominus}/V$
$BrO_3^- + 6H^+ + 5e^- \rightleftharpoons \frac{1}{2}Br_2 + 3H_2O$	+1.52
$\frac{1}{2}I_2 + 2e^- \rightleftharpoons I^-$	+0.54
$Fe^{3+} + e^- \rightleftharpoons Fe^{2+}$	+0.77
$Ag^+ + e^- \rightleftharpoons Ag$	+0.80

Predict, giving reasons, what reaction (if any) could occur in *each* of the following:

(i) Aqueous potassium iodide is added to acidified potassium bromate(V).

....

....

(ii) Metallic silver is added to aqueous iron(III) sulphate.

....

....

(iii) Aqueous iron(III) chloride and aqueous potassium iodide are mixed.

....

....

....

....

....

(15 min)

Answers
Electrochemistry and redox

Oxidation and reduction

Checkpoints

1. (a) (i) iron(II) ion
       (ii) iron(III) chloride
       (iii) chlorate(I) ion
   (b) (i) +5
       (ii) +7
   (c) $2Na(s) + 2H_2O(l) \rightarrow 2NaOH(aq) + H_2(g)$

   It is redox because a sodium atom loses an electron to become a sodium ion. The oxidation number of sodium increases from 0 to +1. The oxidation number of hydrogen decreases from +1 to 0.

> **Examiner's secrets**
>
> Always place the sign in front of the oxidation number. If you write +2 it will be read as an oxidation number (or state) of plus two. If you write 2+ it will be read as a charge of two plus. The same goes for –2 and 2–. Make sure you write what you mean and mean what you write.

2. (a) Excess acid is used to ensure that all the manganate(VII) is reduced to manganese(II) ions. If insufficient acid is used some manganate(VII) ions are only reduced as far as the +4 state.

> **Examiner's secrets**
>
> If the solution in your conical flask becomes cloudy brown (colloidal hydrated MnO_2), add more sulphuric acid.

   (b) 'Quantitatively' means in exact measured amounts (moles) of substances according to the equation.
   (c) With a pipette, which delivers an exact volume of liquid (the solution of reductant in this case).

> **Examiner's secrets**
>
> In some experiments you may have to weigh the amount of solid reductant and then transfer it to your conical flask.

   (d) No indicator is needed because a solution containing aqueous manganate(VII) ions is purple (pink when very dilute) but turns colourless when reduced to a solution containing aqueous manganese(II) ions.

Exam question

(a) This is a redox reaction.
   Reduction
   $$2H^+(aq) + 2e^- \rightarrow H_2(g)$$
   Oxidation
   $$Mg(s) \rightarrow Mg^{2+}(aq) + 2e^-$$
(b) This is an acid–base (proton-transfer) reaction:
   $$2CrO_4^{2-}(aq) + 2H^+(aq) \rightleftharpoons Cr_2O_7^{2-}(aq) + H_2O(l)$$
   It is *not* a redox reaction because the oxidation state of chromium (+6) does not change. The chromate(VI) ion is a Brønsted–Lowry base.

Electrochemistry

Checkpoint

(a) So that there is electrical contact (by the ions present) between the two solutions.
(b) The porous plug prevents rapid mixing of the two solutions.

Exam question

(a) (i) $H_2(g) \rightarrow 2H^+(aq) + 2e^-$
       or $H_2(g) + 2H_2O(l) \rightarrow 2H_3O^+(aq) + 2e^-$
   (ii) $Cu^{2+}(aq) + 2e^- \rightarrow Cu(s)$
(b) $H_2(g) + Cu^{2+}(aq) \rightarrow 2H^+(aq) + Cu(s)$
   or $H_2(g) + 2H_2O(l) + Cu^{2+}(aq) \rightarrow 2H_3O^+(aq) + Cu(s)$

Electrochemical series

Checkpoint

(a) (i) Colourless.
   (ii) Aqueous potassium bromide goes brown due to liberated bromine. Aqueous potassium iodide first goes brown due to the formation of iodine which reacts with iodide ions to give the brown $I_3^-(aq)$ ion. If the chlorine is in excess then a black precipitate of iodine is seen.
(b) $Br_2(aq) + 2KI(aq) \rightarrow 2KBr(aq) + I_2(aq)$

Exam question

(a) $Br_2(aq) + H_2(g) \rightarrow 2Br^-(g) + 2H^+(aq)$
   Br^2 is a stronger oxidant than H^+ and H_2 is a stronger reducer than Br^-
(b) no reaction
   Fe and I^- are both reductants
(c) no reaction
   H_3O^+ and Cl_2 are both oxidants
(d) no reaction
   Cl_2 and Cu^{2+} are both oxidants
(e) $Cu(s) + Br_2(aq) \rightarrow Cu^{2+}(aq) + 2Br^-(aq)$
   Br_2 is an oxidant and Cu is a reductant

Structured exam question

(a) The e.m.f. of an electrochemical cell under standard conditions of temperature (298 K), pressure (1 atm) and electrolyte concentration (1 mol dm^{-3}) represented by the cell diagram

$$Pt[H_2(g)]\,|\,2H^+(aq) \;\vdots\; M^{n+}(aq)\,|\,M(s)$$

   where M(s) and M^{n+}(aq) represent the metal and its cations.

> **Examiner's secrets**
>
> You need to give the standard conditions (T, p and concentration) and mention the standard hydrogen electrode as the left-hand reference electrode.

(b) (i)

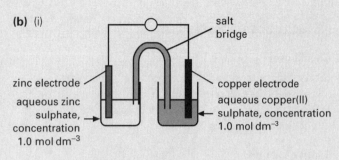

zinc electrode

aqueous zinc sulphate, concentration 1.0 mol dm⁻³

salt bridge

copper electrode

aqueous copper(II) sulphate, concentration 1.0 mol dm⁻³

(ii) E.m.f. is (+0.34) − (−0.76) = +1.10 V.

(iii) A very high internal resistance voltmeter.

(iv) From the zinc to the copper.

(v) $Zn(s) + Cu^{2+}(aq) \rightarrow Zn^{2+}(aq) + Cu(s)$.

	Ion–electron half-equation		$E^{\ominus}/V$
oxidizing power ↓	$\frac{1}{2}I_2 + 2e^- \rightleftharpoons I^-$	reducing power ↑	+0.54
	$Fe^{3+} + e^- \rightleftharpoons Fe^{2+}$		+0.77
	$Ag^+ + e^- \rightleftharpoons Ag$		+0.80
	$BrO_3^- + 6H^+ + 5e^- \rightleftharpoons \frac{1}{2}Br_2 + 3H_2O$		+1.52

(c) (i) Bromate(V) ions would oxidize iodide ions to iodine molecules and be reduced to bromine molecules, $BrO_3^- + 6H^+ + 5I^- \rightarrow \frac{1}{2}Br_2 + 3H_2O + 2\frac{1}{2}I_2$, because the E for the acidified bromate is more positive than that for iodine.

(ii) No reaction because the E for silver ions is more positive than that for iron(III) ions. (Silver ions would oxidize iron(II) ions to iron(III) ions and be reduced to silver.)

(iii) iron(III) ions would oxidize iodide ions to iodine molecules and be reduced to iron(II) ions, $Fe^{3+} + I^- \rightarrow Fe^{2+} + \frac{1}{2}I_2$, because the E for iron(III) ions is more positive than that for iodine.

You must get used to the periodic table but there is no need to learn it by heart. See how elements are organized into groups with characteristic properties. Notice the trends in properties down the table (within each group) and across the table (from group I to group VII). Between groups II and III you should see that the series of d-block elements across the table shows more similarities than trends. Do make sure you check to see which aspects of inorganic chemistry are included in your particular syllabus.

Exam themes

→ Describe and explain the trend in the following properties of elements going across a period: 1st ionization energy, atomic radius, melting point, electronegativity

→ Describe the bonding and reactions of the chlorides or the oxides across a period and the acid–base nature of the oxides across a period

→ Describe and explain the characteristics, reactions and trends in reactivity of groups I to VIII (or 0) and the transition metals

Topic checklist

O AS ● A2	AQA	CCEA	EDEXCEL	OCR	WJEC
The periodic table	O	O	O	O	O
Patterns and trends in the periodic table	O	O	O	O	O
Groups I and II: alkali and alkaline earth metals	O	O	O	O	O
Industrial chemistry: s-block	●	●	●	●	●
Group III: aluminium and boron	●	●	●	●	●
Group IV: elements and oxides	●	●	●	●	●
Group IV: chlorides and hydrides	●	●	●	●	●
Group V: elements and oxides	●	●	●	●	●
Group V: oxoacids and hydrides	●	●	●	●	●
Group VI: oxygen and sulphur	●	●	●	●	●
Group VI: water and hydrogen peroxide	●	●	●	●	●
Group VII: halogens and hydrogen halides	O	O	O	O	O
Group VII: halides and interhalogen compounds	O	O	O	O	O
Industrial chemistry: aluminium and carbon	O●	O●	O●	O●	O●
Industrial chemistry: silicon and nitrogen	O	O	O	O	O
Industrial chemistry: sulphur and the halogens	O	O	O	O	O●
Group VIII (or 0): the noble gases	●	●	●	●	●
1st transition series: metals, aqueous ions and redox	●	●	●	●	●
1st transition series: redox reactions and complex ion formation	●	●	●	●	●
1st transition series: chromium	●	●	●	●	●
1st transition series: manganese	●	●	●	●	●
1st transition series: copper	●	●	●	●	●

The periodic table

Checkpoint

Draw on the periodic table a line which divides metals from non-metals.

s-block

I	II								
1.01 **H** Hydrogen 1									

d-block →

I	II								
6.94 **Li** Lithium 3	9.01 **Be** Beryllium 4								
23.0 **Na** Sodium 11	24.3 **Mg** Magnesium 12								
39.1 **K** Potassium 19	40.1 **Ca** Calcium 20	45.0 **Sc** Scandium 21	47.9 **Ti** Titanium 22	50.9 **V** Vanadium 23	52.0 **Cr** Chromium 24	54.9 **Mn** Manganese 25	55.8 **Fe** Iron 26	58.9 **Co** Cobalt 27	
85.5 **Rb** Rubidium 37	87.6 **Sr** Strontium 38	88.9 **Y** Yttrium 39	92.9 **Zr** Zirconium 40	92.9 **Nb** Niobium 41	95.9 **Mo** Molybdenum 42	98.9 **Tc** Technetium 43	101.1 **Ru** Ruthenium 44	103 **Rh** Rhodium 45	
133 **Cs** Caesium 55	137 **Ba** Barium 56	139 **La** Lanthanum 57	179 **Hf** Hafnium 72	181 **Ta** Tantalum 73	184 **W** Tungsten 74	186 **Re** Rhenium 75	190 **Os** Osmium 76	192 **Ir** Iridium 77	
(223) **Fr** Francium 87	(226) **Ra** Radium 88	(227) **Ac** Actinium 24							

Action point

(a) Describe how the acid–base nature of oxides varies across the period Na to Cl.

(b) State how the nature of the bonding in the chlorides varies from Na to S and illustrate this by describing how at least *one* chloride for *each* element behaves when treated with water.

(c) Compare and explain the difference in the reaction of CCl_4 and $SiCl_4$ with water.

← f-block

140 **Ce** Cerium 58	141 **Pr** Praseodymium 59	144 **Nd** Neodymium 60	(147) **Pm** Promethium 61	150 **Sm** Samarium 62	(153) **Eu** Europium 63
232.0 **Th** Thorium 90	(231) **Pa** Protoactinium 91	238.1 **U** Uranium 92	(237) **NP** Neptunium 93	(244) **Pu** Plutonium 94	(243) **Am** Americium 95

Key

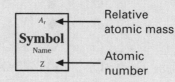

A_r → Relative atomic mass

Symbol Name

Z → Atomic number

p-block

	III	IV	V	VI	VII	4.00 **He** Helium 2
	10.8 **B** Boron 5	12.0 **C** Carbon 6	14.0 **N** Nitrogen 7	16.0 **O** Oxygen 8	19.0 **F** Fluorine 9	20.2 **Ne** Neon 10
→	27.0 **Al** Aluminium 13	28.1 **Si** Silicon 14	31.0 **P** Phosphorus 15	32.1 **S** Sulphur 16	35.5 **Cl** Chlorine 17	40.0 **Ar** Argon 18

58.7 **Ni** Nickel 28	63.5 **Cu** Copper 29	65.4 **Zn** Zinc 30	69.7 **Ga** Gallium 31	72.6 **Ge** Germanium 32	74.9 **As** Arsenic 33	79.0 **Se** Selenium 34	79.9 **Br** Bromine 35	83.8 **Kr** Krypton 36
106 **Pd** Palladium 46	108 **Ag** Silver 47	112 **Cd** Cadmium 48	115 **In** Indium 49	119 **Sn** Tin 50	122 **Sb** Antimony 51	128 **Te** Tellurium 52	127 **I** Iodine 53	131 **Xe** Xenon 54
195 **Pt** Platinum 78	197 **Au** Gold 79	201 **Hg** Mercury 80	204 **Tl** Thallium 81	207 **Pb** Lead 82	209 **Bi** Bismuth 83	(210) **Po** Polonium 84	(210) **At** Astatine 85	(222) **Rn** Radon 86

→

157 **Gd** Gadolinium 64	159 **Tb** Terbium 65	162 **Dy** Dysprosium 66	165 **Ho** Holmium 67	167 **Er** Erbium 68	169 **Tm** Thulium 69	173 **Yb** Ytterbium 70	175 **Lu** Lutetium 71
(247) **Cm** Curium 96	(245) **Bk** Berkelium 97	(251) **Cf** Californium 98	(254) **Es** Einsteinium 99	(253) **Fm** Fermium 100	(256) **Md** Mendelevium 101	(254) **No** Nobelium 102	(257) **Lr** Lawrencium 103

Action point

Show on the periodic table the most electronegative element and the least electronegative element. Indicate which of the elements form compounds which show hydrogen bonding.

Check the net

You can download a copy of the periodic table for your own use from www.webelements.com.

You will need to have Adobe Acrobat Reader 4 installed.
This can be downloaded free of charge from www.adobe.com.

Patterns and trends in the periodic table

"The elements, if arranged according to their atomic weights, exhibit an evident periodicity of properties."

Dmitri Ivanovich Mendeléev

In 1869 Dmitri Mendeléev arranged elements in a table and predicted the existence and properties of two elements that were later discovered. Since then a periodic table in one form or another has been a cornerstone of chemistry.

Periodicity

→ Similar properties recur at regular intervals when the elements are arranged in order of increasing atomic number.
→ Similarity of properties of the elements and their compounds occur within each group of the s- and p-blocks.

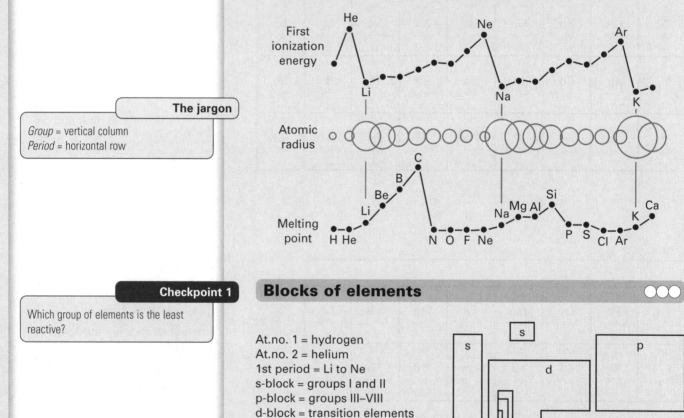

The jargon

Group = vertical column
Period = horizontal row

Checkpoint 1

Which group of elements is the least reactive?

Blocks of elements

At.no. 1 = hydrogen
At.no. 2 = helium
1st period = Li to Ne
s-block = groups I and II
p-block = groups III–VIII
d-block = transition elements
f-block = lanthanides and actinides

→ Most elements are metals.
→ Group I elements are the most reactive metals.
→ Group VII elements are the most reactive non-metals.

Watch out!

Make sure you can relate atypical properties to the following:
→ very small atom size
→ very high polarizing power of the cation
→ very low polarizability of the anion
→ high electronegativity
→ inability of the second shell to extend the number of its electrons beyond eight

Atypical properties

Each element in the first period (Li to F) has some properties not typical of their group.

Group I Decomposition of $Li_2CO_3 \rightarrow Li_2O + CO_2$ on heating in a test tube. The other alkali metal carbonates do not decompose.

Group IV Tetrachloromethane does not hydrolyse in water. The other tetrachloro compounds do, e.g. $SiCl_4 + 2H_2O \rightarrow SiO_2 + 4HCl$.

Group VII Silver fluoride dissolves in water but the other silver halides are insoluble, e.g. $Ag^+(aq) + Cl^-(aq) \rightarrow AgCl(s)$.

Links

See page 89: lithium; 92: boron; 106: fluorine.

Trends in properties

You need to know the important broad trends in the properties of elements and their compounds.

- reactivity of the s-block metals *increases* with increasing atomic number down each group
- reactivity of the halogens and group VI elements *decreases* with increasing atomic number down each group
- character of group IV elements changes from non-metal to metal with increasing atomic number down the group

From left to right across the s- and p-block elements

- atomic and ionic radius decreases
- tendency of chlorides to hydrolyse increases

Within the series of transition elements

- metallic character shows little or no change with increasing atomic number across the series
- basic character decreases and acidic character increases as the oxidation number increases

+2 basic	+3 amphoteric	+6 acidic
Fe^{2+} Cr^{2+}	Fe^{3+} Cr^{3+}	
	FeO_2^- $[Cr(OH)]_6^{3-}$	FeO_4^{2-} CrO_4^{2-} $Cr_2O_7^{2-}$

Some important patterns and trends for the s- and p-blocks are summarized below:

Increase →

Increase ↓	Increase ↑
Elements	*Elements*
Atomic radius	First ionization energy
Ionic radius	Electronegativity
Reducing power	Oxidizing power
Oxides	*Oxides*
Basic character	Acidic character
Oxides, hydrides, chlorides	*Oxides, hydrides, chlorides*
Ionic character	Covalent character

The jargon

Trend means how properties vary with atomic number or position in the periodic table.

Examiner's secrets

Learn these trends and make sure you can connect them to the electronic configurations of the atoms. You will usually be asked to describe and explain trends down groups and across periods.

The jargon

Basic elements: metals – form cations and their (hydr)oxides react with acids.
Acidic elements: non-metals – form anions and their (hydr)oxides react with alkalis.
Amphoteric elements: metals or metalloids – form cations and anions and their (hydr)oxides react with acids and alkalis.

Checkpoint 2

Write an ionic equation for the reaction of
(a) magnesium oxide with aqueous acid
(b) carbon dioxide with aqueous alkali
(c) aluminium oxide with (i) aqueous acid and (ii) aqueous alkali

Exam question (5 min) answer: page 129

(a) Give one example of each of the following
  (i) an amphoteric oxide
  (ii) a semiconductor element
  (iii) a group IV oxide which is an oxidizing agent

(b) Give *two* trends in chemical properties, with increasing atomic number, for the elements of group IV.

Groups I and II: alkali and alkaline earth metals

The Group I elements are the most reactive metals in the Periodic Table. The Group II elements are less reactive and, except for beryllium, more like typical metals.

The elements ●●●

Trends in physical properties

	Electronic configuration	M.p./°C	ρ/g cm^{-3}	E_{m1}/kJ mol^{-1}	
Li	$1s^2 2s$	180	0.53	519	
Na	$1s^2 2s^2 2p^6 3s$	97.8	0.97	494	
K	$1s^2 2s^2 2p^6 3s^2 3p^6 4s$	63.7	0.86	418	
Rb	$1s^2 2s^2 2p^6 3s^2 3p^6 3d^{10} 4s^2 4p^6 5s$	39.9	1.53	402	
Cs	[krypton core]$4d^{10} 5s^2 5p^6 6s$	28.7	1.54	376	
Be	$1s^2 2s^2$	1 278	1.85	900	hcp
Mg	$1s^2 2s^2 2p^6 3s^2$	649	1.74	736	hcp
Ca	$1s^2 2s^2 2p^6 3s^2 3p^6 4s^2$	839	1.54	590	fcc
Sr	$1s^2 2s^2 2p^6 3s^2 3p^6 3d^{10} 4s^2 4p^6 5s^2$	769	2.62	548	fcc
Ba	[krypton core]$4d^{10} 5s^2 5p^6 6s^2$	725	3.51	502	bcc

→ All alkali metals are soft and have a body-centred cubic structure.
→ All s-block elements are ductile, malleable and conduct electricity.

Trends in chemical properties

→ All alkali metals react with water with increasing violence down the group (Li–Cs), e.g. $2Na(s) + 2H_2O(l) \rightarrow 2NaOH(aq) + H_2(g)$.
→ Magnesium burns in water vapour to form magnesium oxide and the rest of the group (Ca–Ba) react with water to form hydroxides.
→ s-Block elements combine with halogens and oxygen to form halides, oxides, peroxides and superoxides.
→ Magnesium reduces $H^+(aq)$ ion in aqueous acids to $H_2(g)$: the other s-block elements react similarly but too violently for safety.
→ Magnesium reduces the nitrate ion in nitric acid and its aqueous salts to the ammonium ion.

The compounds ●●●

Oxides

→ All the oxides are basic except for beryllium oxide which is amphoteric and shows a diagonal relationship with aluminium.
→ Peroxides of sodium (Na_2O_2) and barium (BaO_2) react with water and give off oxygen: $Na_2O_2(s) + H_2O(l) \rightarrow 2NaOH(aq) + \frac{1}{2}O_2(g)$.
→ Potassium, rubidium and caesium form superoxides (KO_2).

Hydroxides

→ Hydroxides form when s-block metals or oxides react with water.
→ All group I hydroxides are water soluble.
→ In group II solubility increases from $Mg(OH)_2$ to $Ba(OH)_2$.

Group II hydroxides are only sparingly soluble in water. Saturated aqueous calcium hydroxide is the 'limewater' which turns milky when you test carbon dioxide: $CO_2(g) + Ca(OH)_2(aq) \rightarrow CaCO_3(s) + H_2O(l)$. The white precipitate 'redissolves' when you pass excess gas into the suspension: $CaCO_3(s) + H_2O(l) + CO_2(g) \rightarrow Ca^{2+}(aq) + 2HCO_3^-(aq)$.

Carbonates

→ Sodium, potassium, rubidium and caesium carbonates are thermally stable and do not decompose on heating in a test tube.
→ You can decompose the carbonates of lithium and the group II metals by heating the solid in a test tube.
→ Thermal stability of group II carbonates increases down the group.
→ Sodium and potassium hydrogencarbonates can exist as solids but calcium hydrogencarbonate exists only in solution.

Nitrates

→ Nitrates are formed when (hydr)oxides react with dilute nitric acid.
→ s-Block nitrates decompose on heating to give off oxygen but lithium and the group II nitrates also give off nitrogen dioxide:

$$2KNO_3(s) \rightarrow 2KNO_2(s) + O_2(g)$$
$$2Ca(NO_3)_2(s) \rightarrow 2CaO(s) + 4NO_2(g) + O_2(g)$$

→ All nitrates are soluble in water.

Sulphates

Calcium sulphate occurs naturally as $CaSO_4$, *anhydrite*, and $CaSO_4 \cdot 2H_2O$, *gypsum*. Magnesium sulphate (Epsom salts) is soluble and crystallizes as $MgSO_4 \cdot 7H_2O$.

The solubility of group II sulphates decreases down the group.

$Ba^{2+}(aq)$ is extremely toxic but $BaSO_4(s)$ is so insoluble that hospital patients swallow a suspension in water for a 'barium meal' X-ray.

Chlorides

→ All chlorides can be formed by direct combination of the elements.
→ All the s-block chlorides are soluble.

Sodium chloride is a major industrial chemical used in the manufacture of sodium hydroxide and chlorine.

Hydrides and nitrides

→ s-Block elements combine with hydrogen to form ionic hydrides.
→ s-Block hydrides react with water to form an alkali and hydrogen.
→ Lithium and magnesium burn in nitrogen to form nitrides.

Checkpoint 2

(a) What is (i) hard water and (ii) limescale?
(b) How do stalactites and stalagmites form?
(c) How might the thermal instability of $CaCO_3$ and Li_2CO_3 be related to the polarizing power of the cation?

Examiner's secrets

We often set questions about trends in group II. We are very fond of the trend in the solubilities of the sulphates and hydroxides and the trend in the thermal stabilities of the carbonates.

Don't forget!

We use aqueous barium chloride containing hydrochloric acid to test for sulphate ions:
$Ba^{2+}(aq) + SO_4^{2-}(aq) \rightarrow BaSO_4(s)$ white ppt.

Links

See pages 108–9: halides.

Links

See page 15: shapes of molecules.

Exam question (6 min)   answer: page 129

Write an equation for the reaction of (a) water with (i) calcium oxide, (ii) barium peroxide, (iii) lithium hydride; (b) magnesium with (i) nitrogen, (ii) titanium(IV) chloride, (iii) aqueous nitrate ions.

Industrial chemistry:
s-block

The chemical industry is vital to our economy. Banks have measured prosperity by the tonnage of sulphuric acid manufactured.

The jargon

Brine is concentrated NaCl(aq).
The *diaphragm* is porous asbestos.
The *cell liquor* is NaCl(aq) and NaOH(aq).

Checkpoint 1

(a) At which electrodes does (i) oxidation occur, (ii) reduction occur? (b) Explain the function of the diaphragm. (c) Suggest why, other than cost, membrane cells are beginning to replace the Castner–Kellner and the diaphragm cells.

Manufacture of sodium hydroxide and chlorine

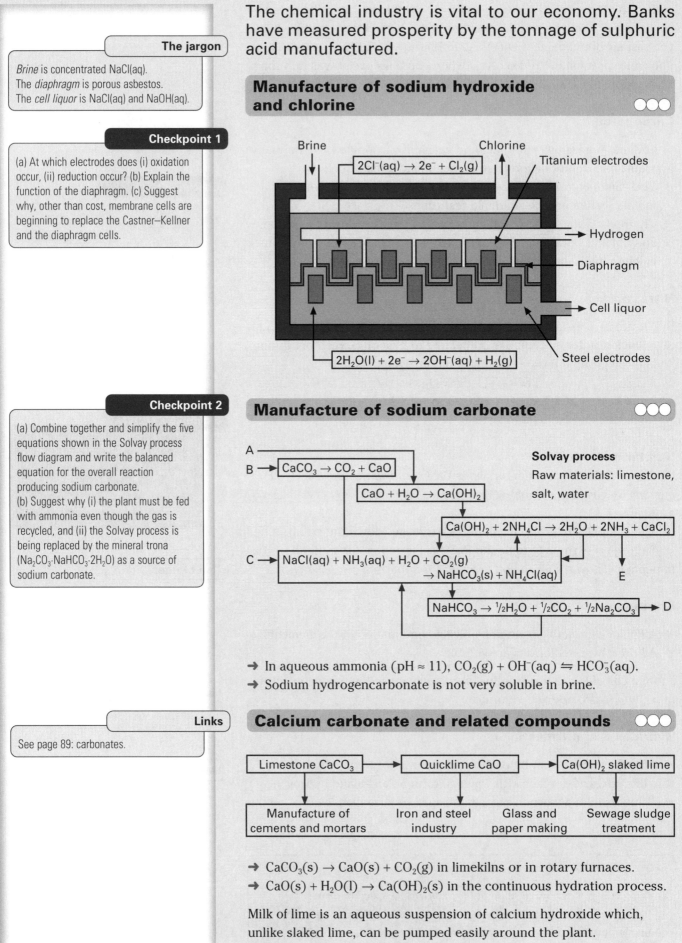

Brine

Chlorine

$2Cl^-(aq) \rightarrow 2e^- + Cl_2(g)$

Titanium electrodes

Hydrogen

Diaphragm

Cell liquor

$2H_2O(l) + 2e^- \rightarrow 2OH^-(aq) + H_2(g)$

Steel electrodes

Checkpoint 2

(a) Combine together and simplify the five equations shown in the Solvay process flow diagram and write the balanced equation for the overall reaction producing sodium carbonate.
(b) Suggest why (i) the plant must be fed with ammonia even though the gas is recycled, and (ii) the Solvay process is being replaced by the mineral trona ($Na_2CO_3 \cdot NaHCO_3 \cdot 2H_2O$) as a source of sodium carbonate.

Manufacture of sodium carbonate

A

B — $CaCO_3 \rightarrow CO_2 + CaO$

$CaO + H_2O \rightarrow Ca(OH)_2$

$Ca(OH)_2 + 2NH_4Cl \rightarrow 2H_2O + 2NH_3 + CaCl_2$

C — $NaCl(aq) + NH_3(aq) + H_2O + CO_2(g) \rightarrow NaHCO_3(s) + NH_4Cl(aq)$

E

$NaHCO_3 \rightarrow \frac{1}{2}H_2O + \frac{1}{2}CO_2 + \frac{1}{2}Na_2CO_3$ → D

Solvay process
Raw materials: limestone, salt, water

→ In aqueous ammonia (pH ≈ 11), $CO_2(g) + OH^-(aq) \leftrightarrows HCO_3^-(aq)$.
→ Sodium hydrogencarbonate is not very soluble in brine.

Links

See page 89: carbonates.

Calcium carbonate and related compounds

Limestone $CaCO_3$	→	Quicklime CaO	→	$Ca(OH)_2$ slaked lime

Manufacture of cements and mortars	Iron and steel industry	Glass and paper making	Sewage sludge treatment

→ $CaCO_3(s) \rightarrow CaO(s) + CO_2(g)$ in limekilns or in rotary furnaces.
→ $CaO(s) + H_2O(l) \rightarrow Ca(OH)_2(s)$ in the continuous hydration process.

Milk of lime is an aqueous suspension of calcium hydroxide which, unlike slaked lime, can be pumped easily around the plant.

Extraction of s-block elements ●●●

→ s-Block metals have been extracted by electrolysis of molten but *not* aqueous electrolytes.
→ Calcium and magnesium can be extracted from carbonate ores by non-electrolytic methods.

Sodium

at anode:
$2Cl^- \rightarrow Cl_2 + 2e^-$
at cathode:
$2Na^+ + 2e^- \rightarrow 2Na$

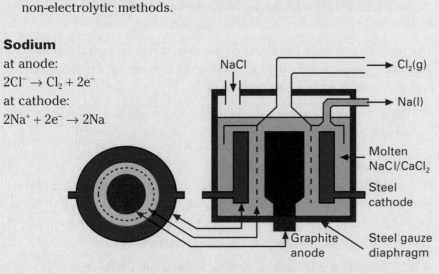

Calcium chloride lowers the melting point of the sodium chloride.

Calcium

The electrolytic extraction of calcium from molten calcium chloride has been replaced by its manufacture from limestone.

1 Limestone is turned into quicklime.
2 Aluminium powder is mixed with quicklime.
3 Briquettes of the mixture are heated under reduced pressure:

$$6CaO(s) + 2Al(s) \rightarrow Ca_3Al_2O_3(s) + 3Ca(g)$$

Magnesium

From seawater:

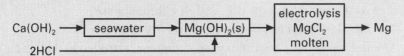

From magnesite (MgO) or dolomite ($CaCO_3 \cdot MgCO_3$):

1 Dolomite is converted into oxide.
2 Iron/silicon alloy is mixed with the oxide.
3 The mixture is heated under reduced pressure to give magnesium vapour.

Exam question (10 min)   answer: page 129

(a) State and explain, with the help of equations, what is observed when (i) a piece of magnesium is heated in air, (ii) one drop of water is added to the residue in a test tube, (iii) a piece of damp red litmus paper is held in the mouth of the test tube.

(b) State one way in which (i) lithium resembles magnesium; (ii) beryllium resembles aluminium.

Checkpoint 3

Suggest why
(a) the anode is made of graphite and not steel
(b) the anode is surrounded by a steel gauze diaphragm
(c) the sodium is not pure

Checkpoint 4

(a) Write an equation for the conversion of limestone into quicklime. (b) What is (i) the function of the aluminium in this reaction, (ii) the oxidation number of the calcium and aluminium before and after reaction?

Group III: aluminium and boron

Aluminium is the most important element of this first group in the p-block. For A-level, boron is unimportant.

Boron ●●●

→ Boron *never* forms B^{3+} ions under ordinary conditions.
→ Boron is non-metallic and forms covalent compounds.

The tendency to form covalent compounds is related to the cation polarizing power which increases as the charge on the ion increases and as the radius of the ion decreases.

→ As the first element in group III boron shows atypical properties as the nuclear charge of its small atom is almost unshielded.

Electron-deficient compounds

BCl_3 and BF_3 are covalent compounds in which the valence shell of the boron atom has only six electrons (*not* eight). Each molecule has three bonding pairs of electrons only but no lone (non-bonding) pair. Such compounds can act as Lewis acids and are called *electron deficient*.

→ BCl_3 and BF_3 are trigonal planar molecules and Lewis acids.
→ Ammonia reacts with boron trihalide in a Lewis acid–base reaction with the formation of a dative covalent bond:

$$H_3N\text{:} + BCl_3 \rightarrow H_3N-BCl_3$$

Sodium tetrahydridoborate(III)

You may meet this compound in organic chemistry questions. Its traditional name is sodium borohydride. $Na^+ BH_4^-$ is not as reactive as its aluminium counterpart so we can use it in aqueous solution to reduce aldehydes and ketones.

Aluminium ●●●

→ Aluminium is a highly reactive metal protected by a tough layer of the oxide preventing reaction with the atmosphere.
→ The metal is extracted from *bauxite* ore by electrolysis.
→ Aluminium combines with oxygen, halogens, nitrogen and sulphur at high temperature or when the oxide layer is removed.
→ The oxidation number of aluminium in its compounds is +3.

Aluminium compounds ●●●

The high charge and small size of the cation make Al^{3+} strongly polarizing so that many aluminium compounds are covalent.

Aluminium chloride

When dry chlorine is passed over heated aluminium, anhydrous aluminium chloride forms as a covalent sublimate:

$$2Al(s) + 3Cl_2(g) \rightarrow Al_2Cl_6(s)$$

Measurements of M_r in non-aqueous solvents and in the vapour state show that aluminium chloride forms a dimer whose structure can be explained in terms of the electron deficiency of the monomer, $AlCl_3$.

Checkpoint 1

Write the ground state electronic configurations of a B atom and an Al atom.

Examiner's secrets

This reaction is often used to test your understanding of bonding and shapes of molecules.

Checkpoint 2

Use the VSEPR theory to predict the shape of
(a) NH_3
(b) H_3N-BCl_3
(c) BH_4^-.

The jargon

Sublime means to change state from solid to vapour (gas) without melting. *Sublimate* is the solid obtained when a vapour (gas) condenses without forming a liquid.

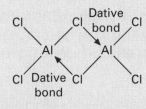

Each Al atom can accept a lone pair of electrons from a Cl atom in the other AlCl$_3$ molecule

Links

See page 14: bonding.

Aluminium oxide

Bauxite, $Al_2O_3 \cdot 2H_2O$, is the mineral from which the metal is extracted by electrolysis. Anhydrous aluminium oxide occurs naturally as corundum and emery. These two very hard minerals are used industrially as abrasives. Ruby, sapphire and topaz are corundum containing chromium, titanium and iron respectively.

→ Aluminium oxide is amphoteric reacting with acids to form $Al(H_2O)_6^{3+}(aq)$ and alkalis to form $Al(OH)_4^-(aq)$.

Aluminium oxide is a refractory material used in furnace linings and in the nose cones of rockets and space vehicles. As a fine powder, it can be used in column **chromatography** and serve as a catalyst in the dehydration of ethanol to ethene.

Watch out!

You may use washing soda to remove tea stains from cups but *never* from aluminium teapots!

Aluminium hydroxide

When you add NaOH(aq) to aluminium salt solutions you can expect a white gelatinous precipitate to form and then disappear. A simple explanation is that aluminium hydroxide precipitates and then redissolves in excess alkali to form sodium aluminate solution. For a more elaborate interpretation you should refer to

1 the polarizing effect of Al^{3+} ion on H_2O molecules making
2 the aqueous cation act as a Brønsted–Lowry acid
   $[Al(H_2O)_6]^{3+} \rightleftharpoons [Al(H_2O)_5(OH)]^{2+} + H^+$ and
3 the shift in the equilibrium positions as the addition of OH^- increases the pH causing successive ionizations to produce
4 the white gelationous precipitate $[Al(H_2O)_3(OH)_3](s)$ and then
5 the colourless solution containing $[Al(OH)_6]^{3-}(aq)$

The jargon

Refractory means able to withstand very high temperatures (>2 000 °C) without melting or decomposing. *Gelatinous precipitate* is a jelly-like precipitate that settles slowly if at all. *Granular precipitate* is a powder-like precipitate that settles quickly.

Checkpoint 3

State and explain what is formed when you mix sodium carbonate, $Na_2CO_3(aq)$, with aqueous aluminium sulphate to produce a white precipitate and a colourless gas.

Aluminium sulphate and alums

→ *Excess* dilute sulphuric acid dissolves Al(s) and $Al_2O_3(s)$ only very slowly but freshly precipitated $[Al(H_2O)_3(OH)_3](s)$ very quickly to give a solution from which $Al_2(SO_4)_3 \cdot 18H_2O(s)$ will crystallize.
→ *Alum* is potassium aluminium sulphate, $KAl(SO_4)_2 \cdot 12H_2O$, a *double salt* that crystallizes from an aqueous solution containing equal amounts of potassium sulphate and aluminium sulphate.
→ *Alums* are isomorphous double sulphates having the formula $M^IM^{III}(SO_4)_2 \cdot 12H_2O$ where M^I is a group I ion or the ammonium ion and M^{III} usually a transition element ion or the aluminium ion.

The jargon

Isomorphous means having the same shape or angles and capable of overgrowth (one crystal can grow on top of another).

Exam question (6 min)   answer: page 130

Compare and explain the effect of adding aqueous ammonia to separate aqueous solutions of aluminium sulphate and zinc sulphate.

Group IV: elements and oxides

These elements show more strikingly than any other p-block group the typical change in character from non-metal to metal with increasing atomic number down the group.

Checkpoint 1

List three properties of metals which distinguish them from non-metals.

The jargon

Inert pair effect is the tendency for two valence electrons to form a lone pair and not take part in bonding as, for example, in tin(II) and lead(II) ions.
(g) = graphite
(d) = diamond
at. radius = covalent or metallic radius of an atom

Checkpoint 2

What is the name of the allotropes represented by structures A, B, C?

Examiner's secrets

You are not expected to know the chemistry of germanium but it could be discussed to test your knowledge and understanding of patterns and trends in group IV chemistry.

The jargon

Tin plating is coating steel with tin.
Galvanizing is coating steel with zinc.

Checkpoint 3

(a) Explain why when scratched and exposed to damp air a tin can rusts but a galvanized plate does not.
(b) Suggest why the demand for and therefore the production of lead is decreasing.

The elements

→ All five can have an oxidation number of +4 in covalent compounds; they do not form a simple ion with a 4+ charge.
→ Stability of the +2 oxidation state increases down the group with tin and lead able to form $Sn^{2+}(aq)$ and $Pb^{2+}(aq)$.

	Ground state electronic structure	Density /g cm^{-3}	M.p. /°C	B.p. /°C	At. radius /nm
C	$1s^2 2s^2 2p^2$	2.26(g) 3.51(d)	3 730 sublimes		0.077
Si	$[Ne]3s^2 3p^2$	2.33	1 410	2 360	0.117
Ge	$[Ar]3d^{10}4s^2 4p^2$	5.32	937	2 830	0.122
Sn	$[Kr]4d^{10}5s^2 5p^2$	7.3	232	2 270	0.140
Pb	$[Xe]4f^{14}5d^{10}6s^2 6p^2$	11.4	327	1 744	0.154

Carbon

→ Carbon is the basis of organic life-forms and the largest group of compounds studied by any one group of chemists.
→ Carbon exists in at least three well-defined allotropic forms.

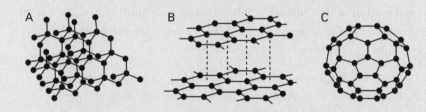

A   B   C

→ Carbon (graphite) fibres are made from acrylic fibre by heating first in air (to 300 °C) and then in an inert gas (to 1 500 °C).

Silicon

→ Silicon is a shiny, blue-grey solid with a diamond structure.
→ Silicon, unlike diamond, is a semiconductor used primarily in the production of microprocessor chips for the computer industry.

Tin

→ Tin has two common allotropes – grey tin (diamond structure) and white tin (metallic structure).
→ Tin is used mainly for plating steel but it is also a constituent of bronze – an alloy employed for thousands of years.

Lead

→ Lead is metallic and used as a roofing material and in shields against radioactive emissions, ionizing radiation and X-rays.
→ Lead is used everyday in lead/acid car batteries and in solders when alloyed with white tin and other metals.

Carbon dioxide and carbon monoxide

→ Carbon dioxide is acidic and is vital in establishing aqueous equilibria that buffer the blood at a constant pH:

$$CO_2(g) + H_2O(l) \rightleftharpoons H_2CO_3(aq) \rightleftharpoons H^+(aq) + HCO_3^-(aq)$$
$$HCO_3^-(aq) \rightleftharpoons CO_3^{2-}(aq) + H^+(aq)$$

→ Carbon monoxide is neutral, almost insoluble in water and very toxic, displacing oxygen to form carboxyhaemoglobin in blood.
→ Carbon monoxide will reduce metal oxides and burn in air:

$$CO(g) + PbO(s) \rightarrow CO_2(g) + Pb(s)$$

→ O=C=O is non-polar and :C≡O: is slightly polar with the oxygen atom being the *positive* end.

:CO and :NO are isoelectronic. The lone pair of electrons (on C and N) makes the molecules powerful ligands capable of forming complexes. For example, the reaction $Ni(s) + 4CO(g) \rightleftharpoons Ni(CO)_4(l)$ is the basis of the Mond process for extracting nickel from its ores or recovering the metal from scrap.

→ Carbon dioxide is readily identified by the limewater test:

$$CO_2(g) + Ca(OH)_2(aq) \rightarrow CaCO_3(s) + H_2O(l)$$

The jargon

Isoelectronic structures have the same number of electrons arranged in the same way, e.g. CH_4 and NH_4^+.

Silicon dioxide

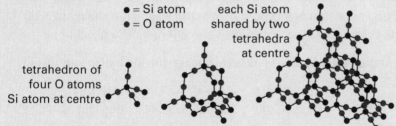

● = Si atom
● = O atom

each Si atom shared by two tetrahedra at centre

tetrahedron of four O atoms
Si atom at centre

→ Silicon dioxide occurs in different forms as a covalent giant molecular structure with the empirical formula SiO_2.
→ SiO_2 is insoluble in water but regarded as an acidic oxide forming silicates when fused with alkalis, e.g. in the blast furnace $CaO + SiO_2 \rightarrow CaSiO_3$ (calcium silicate – slag) or in glass manufacture.

Tin and lead oxides

→ Tin(II) oxide, SnO, is amphoteric and a powerful reducing agent which is spontaneously oxidized by air.
→ Lead(II) oxide, PbO, is amphoteric, exists in yellow or red forms and can be prepared by heating the metal in air.
→ Tin(IV) oxide, SnO_2, is stable and occurs naturally as the ore *cassiterite* from which tin is extracted.
→ Lead(IV) oxide, PbO_2, is insoluble in water, does not react with dilute acids but forms plumbate(IV) ion, $Pb(OH)_6^{2-}$, with alkalis.
→ Lead(IV) oxide reacts as an oxidizing agent with concentrated hydrochloric acid: $PbO_2 + 4HCl \rightarrow PbCl_2 + Cl_2 + 2H_2O$.

Checkpoint 4

How do we account for the use of SiO_2 in (a) furnace linings, (b) abrasives?

The jargon

Hygroscopic means able to absorb moisture from the air.
Deliquescent means able to absorb enough moisture from the air to become an aqueous solution.

Watch out!

Don't confuse silicon, silica and silicone.

Exam question (6 min)   answer: page 130

(a) State and explain what would be observed when excess carbon dioxide is passed into limewater. (b) Explain the structure and shape of the carbonate ion.

Group IV: chlorides and hydrides

The tetrachlorides of group IV are all covalently bonded and their behaviour is dependent on this. But why is it that CCl_4 has no reaction with water whilst the reaction of $SiCl_4$ and water can be quite dramatic? The ways in which Pb^{2+} can be identified in analysis can also be seen here.

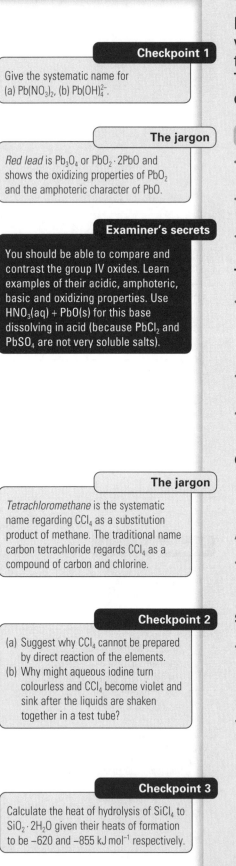

Checkpoint 1

Give the systematic name for
(a) $Pb(NO_3)_2$, (b) $Pb(OH)_4^{2-}$.

The jargon

Red lead is Pb_3O_4 or $PbO_2 \cdot 2PbO$ and shows the oxidizing properties of PbO_2 and the amphoteric character of PbO.

Examiner's secrets

You should be able to compare and contrast the group IV oxides. Learn examples of their acidic, amphoteric, basic and oxidizing properties. Use $HNO_3(aq) + PbO(s)$ for this base dissolving in acid (because $PbCl_2$ and $PbSO_4$ are not very soluble salts).

The jargon

Tetrachloromethane is the systematic name regarding CCl_4 as a substitution product of methane. The traditional name carbon tetrachloride regards CCl_4 as a compound of carbon and chlorine.

Checkpoint 2

(a) Suggest why CCl_4 cannot be prepared by direct reaction of the elements.
(b) Why might aqueous iodine turn colourless and CCl_4 become violet and sink after the liquids are shaken together in a test tube?

Checkpoint 3

Calculate the heat of hydrolysis of $SiCl_4$ to $SiO_2 \cdot 2H_2O$ given their heats of formation to be -620 and -855 kJ mol^{-1} respectively.

Chlorides

➜ Only tin and lead form a chloride in which the element has an oxidation number of +2 and aqueous cation, $Sn^{2+}(aq)$ and $Pb^{2+}(aq)$.
➜ All five elements form a covalent tetrachloride whose thermal stability and resistance to hydrolysis decreases down the group.
➜ All except CCl_4 hydrolyse to the oxide XO_2 and HCl.

Tetrachloromethane

➜ CCl_4 *cannot* be prepared by direct combination of the elements even though the reaction is exothermic and energetically feasible:

$$C(graphite) + 2Cl_2(g) \rightarrow CCl_4(l); \Delta H_{298}^{\ominus} = -129.6 \text{ kJ mol}^{-1}$$

➜ Carbon tetrachloride is a dense, colourless liquid used widely as an industrial solvent even though it is carcinogenic and toxic.
➜ CCl_4 is *not* attacked even by hot water in spite of the hydrolysis theoretically being exothermic and energetically feasible.

Calculating the standard enthalpy change for the hydrolysis of CCl_4:

$$CCl_4(l) + 2H_2O(l) \rightarrow CO_2(g) + 4HCl(g)$$
$$\Delta H_{f,298}^{\ominus} \quad -130 \qquad 2 \times (-286) \quad -394 \quad 4 \times (-92)$$

$\Delta H_{298}^{\ominus}$ is $[(-394) + 4 \times (-92)] - [(-130) + 2 \times (-286)] = -60$ kJ mol^{-1}.

➜ Hydrolysis of CCl_4 is kinetically hindered because the C-atom *cannot* extend its valence electron shell beyond eight.

Silicon tetrachloride

➜ $SiCl_4$ is a colourless liquid which fumes in moist air and hydrolyses violently on contact with water:

$$SiCl_4(l) + 4H_2O(l) \rightarrow SiO_2 \cdot 2H_2O(s) + 4HCl(g)$$

➜ Hydrolysis of $SiCl_4$ is *not* kinetically hindered because the Si-atom *can* extend its valence shell beyond the octet of (eight) electrons.

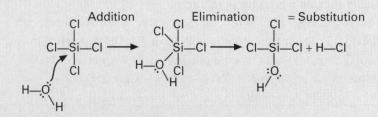

→ Energy released by formation of the strong H_2O—$SiCl_4$ bond provides the energy absorbed by breaking the weak Si—Cl bond.
→ For CCl_4 a strong C—Cl must break before a C—O can form, so the activation energy is very high and the reaction extremely slow.

The products of the hydrolysis would be attacked until all the Cl-atoms have been replaced (substituted) by OH groups.

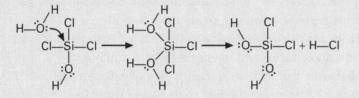

Elimination of water from $Si(OH)_4$ could produce SiO_2.

Tin(IV) chloride and lead(IV) chloride

→ These are covalent liquids hydrolysed by water and moist air.
→ Both will dissolve in concentrated hydrochloric acid to form complex hexachloroanions $[SnCl_6]^{2-}$ and $[PbCl_6]^{2-}$.

Tin(II) chloride and lead(II) chloride

→ Both chlorides are regarded as ionic with some covalent character.
→ $SnCl_2(aq)$ is dissolved in $HCl(aq)$ to prevent hydrolysis and kept in contact with metallic tin to prevent oxidation to the tin(IV) state.
→ $PbCl_2(aq)$ contains $Pb^{2+}(aq)$ but $PbCl_2(s)$ precipitates when a hot solution cools because lead(II) chloride is a sparingly soluble salt.
→ Tin(II) chloride is a strong reducing agent:

$$SnCl_2 + 2HgCl_2 \rightarrow SnCl_4 + Hg_2Cl_2 \text{ (white precipitate)}$$
$$Hg_2Cl_2 + SnCl_2 \rightarrow SnCl_4 + 2Hg \text{ (grey turning to silver liquid)}$$

Hydrides ●●●

All five elements form covalent tetrahydrides but the stability decreases down the group with PbH_4 formed only in traces.
Si forms silanes, Si_nH_{2n+2}, but all hydrolyse even in moist air.

Examiner's secrets

Candidates are often asked to explain why $SiCl_4$ is hydrolysed by water but CCl_4 is not. Remember the Li–F valence shell maximum is 8 electrons, *no 3d*, but the Na–Cl valence shell can extend >8.

Checkpoint 4

What is the systematic name for the $[SnCl_6]^{2-}$ and $[PbCl_6]^{2-}$ anions?

Exam question (10 min)   answer: page 130

(a) Give equations and outline how, starting from the element, you would make a sample of (i) silicon tetrachloride, and (ii) lead(II) chloride.

(b) Explain what happens on adding (i) silicon tetrachloride to aqueous sodium hydroxide, and (ii) concentrated hydrochloric acid to lead(II) chloride.

Group V: elements
and oxides

These elements typically change character from non-metal to metal with increasing atomic number down the group.

Watch out!

White phosphorus is extremely dangerous, igniting spontaneously at body temperature and producing very serious burns.

Checkpoint 1

What other gases are obtained from the atmosphere when extracting nitrogen by fractionation of liquid air?

The jargon

Nitrogen fixation is the process of making atmospheric nitrogen available for plants and animals.

Checkpoint 2

If 🌲 is larger than 🌱 in the following diagram of the layer structure of boron nitride, deduce which represents nitrogen.

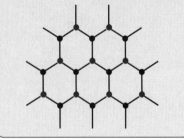

The elements ●●●

→ All five can have oxidation numbers of +5 and +3 in their compounds but the stability of the +5 oxidation state decreases with increasing atomic number down the group from N to Bi.
→ Nitrogen is the only element to triple bond with itself N≡N.
→ White phosphorus and one form of both antimony and arsenic can exist as tetraatomic molecules in which each atom is attached to the other three by single covalent bonds.

	Ground state electronic structure	Density/g cm^{-3}	M.p./°C	B.p./°C
N	$1s^2 2s^2 2p^3$	0.81 at 77 K	−210	
P	$[Ne]3s^2 3p^3$	1.82 white	44.2 white	
		2.34 red	590 red	
As	$[Ar]3d^{10}4s^2 4p^3$	5.72 grey		613 sublimes
Sb	$[Kr]4d^{10}5s^2 5p^3$	6.62	630	1 380
Bi	$[Xe]4f^{14}5d^{10}6s^2 6p^3$	9.80	271	1 560

Nitrogen

→ Nitrogen is about 78% by volume of our atmosphere from which it can be extracted by liquefaction and fractional distillation.
→ Nitrogen exists as very stable diatomic gas molecules and shows some properties atypical of group V.
→ Nitrogen will combine with lithium (and other very electropositive metals) to form the nitride ion, N^{3-}.
→ Nitrogen will combine at high temperatures with non-metals such as boron, hydrogen and oxygen to form BN, NH_3 and oxides.

$2B(s) + N_2(g) \rightarrow 2BN(s)$ (structure similar to graphite)
$3H_2(g) + N_2(g) \rightleftharpoons 2NH_3(g)$ (nitrogen fixation by the Haber process)
$N_2(g) + O_2(g) \rightarrow 2NO(g)$ (lightning or spark discharging in air)

Phosphorus

→ Phosphorus has two allotropes.
→ White phosphorus, P_4, is very reactive and stored under water.
→ Red phosphorus is a less reactive polymeric form of the element.
→ Phosphorus combines directly with oxygen and chlorine forming oxides and chlorides in which its oxidation state can be +3 and +5.

$4P(s) + 3O_2(g) \rightarrow P_4O_6(s); 4P(s) + 5O_2(g) \rightarrow P_4O_{10}(s)$ (excess O_2)
$2P(s) + 3Cl_2(g) \rightarrow 2PCl_3(l); 2P(s) + 5Cl_2(g) \rightarrow [PCl_4]^+[PCl_6]^-(s)$

Arsenic, antimony and bismuth

→ The chemistry of these three elements is not important for A-level except for illustrating the group trend from non-metal to metal.

Oxides

Oxides of nitrogen

→ The oxidation state of nitrogen varies from +1 to +5 in its oxides N_2O, NO, N_2O_3 (very unstable), NO_2 ($\rightleftharpoons N_2O_4$), N_2O_5 (unstable).

The oxides have *delocalized structures* that cannot be represented using only '·', '—' and '→' for an electron, a shared electron and a donated pair of electrons.

$$:N\equiv N \rightarrow \ddot{\underset{..}{O}}: \qquad :\dot{\underset{..}{N}}=\ddot{\underset{..}{O}}:$$

The jargon

Traditional names: *nitrous oxide* (*laughing gas*) = N_2O; *nitric oxide* = NO.

Links

See page 17: delocalized bonding.

→ Dinitrogen oxide (nitrogen(I) oxide) is a colourless gas readily decomposed by a red-hot wooden splint: $2N_2O \rightarrow 2N_2 + O_2$.
→ Nitrogen (mon)oxide, NO, is an odd-electron molecule that reacts spontaneously with oxygen to form NO_2 (brown gas).
→ NO_2 dimerizes to dinitrogen tetraoxide, N_2O_4 (colourless gas).
→ On warming, dinitrogen tetraoxide (colourless) decomposes reversibly into nitrogen dioxide (brown): $N_2O_4(g) \rightleftharpoons 2NO_2(g)$.
→ At 150 °C the gas contains almost 100% of NO_2 molecules and at higher temperatures the nitrogen dioxide decomposes into nitrogen oxide and oxygen: $2NO_2(g) \rightleftharpoons 2NO(g) + O_2(g)$.
→ Dinitrogen pentoxide, $NO_2^+NO_3^-(s)$, decomposes above its melting point in a first order reaction: $N_2O_5(s) \rightarrow 2N_2O_4(g) + O_2(g)$.

Checkpoint 3

Explain why a glowing splint rekindles when put into a test tube of N_2O.

Examiner's secrets

The decomposition of N_2O_4 is often used to test the law of chemical equilibrium and the decomposition of N_2O_5 to test first order kinetics.

Oxides of phosphorus

→ Two solid oxides are produced by burning phosphorus in (limited or excess) air: $4P + 3O_2 \rightarrow P_4O_6$ and $4P + 5O_2 \rightarrow P_4O_{10}$.
→ In the gas phase, the structure of the oxide molecules is based on a tetrahedron of phosphorus atoms.

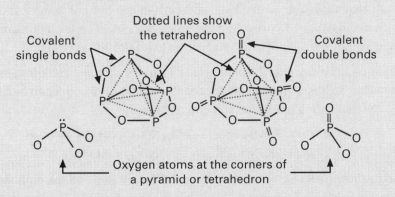

→ Phosphorus(V) oxide reacts violently with water and is used in organic chemistry as a powerful dehydrating agent.

Exam question (10 min)   answer: page 130

(a) Give an equation and explain why (i) nitrogen oxide turns brown in air, and (ii) dinitrogen oxide rekindles a glowing splint.

(b) Give the name and formula of the sodium salt formed when excess sodium hydroxide reacts with (i) N_2O_3, (ii) N_2O_5, (iii) P_4O_6, (iv) P_4O_{10}.

Group V: oxoacids and hydrides

For A-level chemistry, the three most important oxoacids are nitric(V) acid, nitric(III) acid and phosphoric(V) acid.

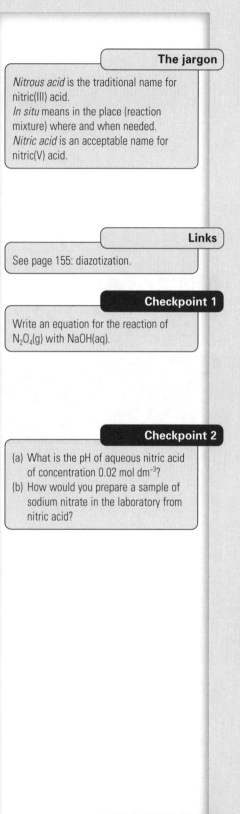

The jargon

Nitrous acid is the traditional name for nitric(III) acid.
In situ means in the place (reaction mixture) where and when needed.
Nitric acid is an acceptable name for nitric(V) acid.

Links

See page 155: diazotization.

Checkpoint 1

Write an equation for the reaction of $N_2O_4(g)$ with NaOH(aq).

Checkpoint 2

(a) What is the pH of aqueous nitric acid of concentration 0.02 mol dm^{-3}?
(b) How would you prepare a sample of sodium nitrate in the laboratory from nitric acid?

Checkpoint 3

What is the conjugate base of H_3PO_4?

The jargon

Phosphoric acid = orthophosphoric acid = tetraoxophosphoric(V) acid.

Nitrous acid ●●●

→ Nitrous acid, HNO_2, disproportionates into nitric acid and nitrogen monoxide and is also rapidly oxidized by air to nitric acid:

$$3HNO_2(aq) \rightarrow HNO_3(aq) + H_2O(l) + 2NO(g)$$
oxidation nos.   +3   +5   +2

→ We usually use ice-cold aqueous sodium nitrite and HCl(aq) to prepare $HNO_2(aq)$ *in situ* for diazotizations in organic chemistry:

$$NaNO_2(aq) + HCl(aq) \rightarrow HNO_2(aq) + NaCl(aq)$$

→ Sodium (or potassium) nitrite is produced when sodium (or potassium) nitrate is heated: $2NaNO_3(s) \rightarrow 2NaNO_2(s) + O_2(g)$.

An aqueous mixture of nitrite and nitrate is produced when nitrogen dioxide (and dinitrogen tetraoxide) reacts with water or alkali.

Nitric acid

→ Nitric acid, HNO_3, can be collected as a yellow distillate when a mixture of sodium nitrate and concentrated sulphuric acid is heated:

$$NaNO_3(s) + H_2SO_4(l) \rightarrow NaHSO_4(s) + HNO_3(l)$$

→ The acid is coloured by nitrogen dioxide formed by thermal decomposition of some nitric acid vapour:

$$2HNO_3(g) \rightarrow H_2O(l) + 2NO_2(g) + \frac{1}{2}O_2(g)$$

→ Nitric acid is manufactured on the industrial scale by the catalytic oxidation of ammonia.

In dilute aqueous solution, nitric acid behaves as a typical strong mono-basic when it reacts with alkalis and carbonates but not when it reacts with metals above hydrogen in the electrochemical series. For example, magnesium reduces $NO_3^-(aq)$ to $NH_4^+(aq)$ and not $H^+(aq)$ to $H_2(g)$:

→ Aqueous nitric acid does *not* react with metals to form hydrogen.
→ Nitric acid is a strong oxidant readily reduced by copper to NO_2 or NO depending upon the acid concentration:

$$Cu + 4HNO_3(conc.) \rightarrow Cu(NO_3)_2 + 2H_2O + 2NO_2$$
$$3Cu + 8HNO_3(dil.) \rightarrow 3Cu(NO_3)_2 + 4H_2O + 2NO$$

All nitrates (salts of nitric acid) are soluble in water and decompose on heating to give off oxygen and (except for the alkali metal nitrates) poisonous brown fumes of nitrogen dioxide:

$$2Cu(NO_3)_2 \rightarrow 2CuO + 4NO_2 + 2O_2$$

Oxoacids of phosphorus

Phosphorus has a larger range of oxoacids than nitrogen in spite of having a smaller range of oxides than nitrogen.

→ P_4O_6 is the anhydride of phosphonic acid, H_3PO_3, and P_4O_{10} is the anhydride of phosphoric acid, H_3PO_4.

Chlorides

Nitrogen forms only the trichloride (not important for A-level) but phosphorus forms the trichloride and the pentachloride by direct reaction with the appropriate amounts of chlorine.

Phosphorus chlorides

→ $PCl_3(l)$ and $PCl_5(s)$ react vigorously with O–H (in water, alcohols, etc.) and N–H (in NH_3 and amines) to give steamy fumes of HCl:

$$2PCl_3 + 3ROH \rightarrow 3RCl + H_3PO_3 + 3HCl$$

→ PCl_3 and gaseous PCl_5 are covalent but *solid* pentachloride has an ionic structure.

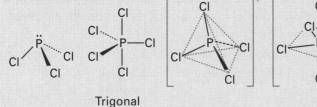

Pyramidal	Trigonal bipyramidal	Tetrahedral	Octahedral

Hydrides

Both elements form two covalent hydrides but only ammonia, NH_3, and hydrazine, N_2H_4, are important for A-level.

Ammonia

→ Ammonia is manufactured from N_2 and H_2 by the Haber process.
→ Ammonia gas (b.p. $-33\,°C$) liquefies easily, dissolves well in water forming a weak alkali and forms ammonium salts with acids:

$$NH_3(aq) + H_2O(l) \rightleftharpoons NH_4^+(aq) + OH^-(aq)$$

→ We use the formation of a white smoke with hydrogen chloride as a test for either gas:

$$NH_3(g) + HCl(g) \rightleftharpoons NH_4Cl(s)$$

→ Ammonia is given off when ammonium salts are heated with s-block alkalis:

$$NH_4^+(aq) + OH^-(aq) \rightarrow NH_3(g) + H_2O(l)$$

→ Ammonium salts readily dissolve in water and dissociate or decompose on heating:

$$NH_4NO_3(s) \rightarrow N_2O(g) + 2H_2O(l)$$
$$NH_4NO_2(aq) \rightarrow N_2(g) + 2H_2O(l)$$

→ Ammonia is an important industrial chemical used mainly for the production of fertilizers, nitric acid (by oxidation) and nylon.

Exam question (6 min)   answer: page 131

(a) Write a balanced equation and state the oxidation number of nitrogen in each product when dinitrogen tetraoxide reacts with aqueous sodium hydroxide.

(b) Calculate (i) the mass of water needed to convert 0.1 mol PCl_3 into H_3PO_3 and (ii) the maximum theoretical volume of HCl(g) produced at room temperature. [molar gas volume = 24 dm^3 at room temperature]

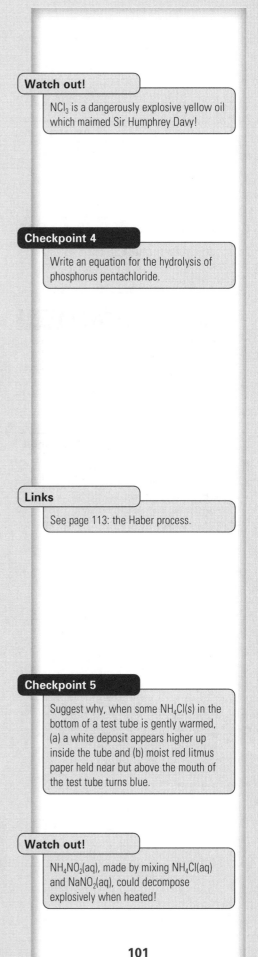

Watch out!

NCl_3 is a dangerously explosive yellow oil which maimed Sir Humphrey Davy!

Checkpoint 4

Write an equation for the hydrolysis of phosphorus pentachloride.

Links

See page 113: the Haber process.

Checkpoint 5

Suggest why, when some $NH_4Cl(s)$ in the bottom of a test tube is gently warmed, (a) a white deposit appears higher up inside the tube and (b) moist red litmus paper held near but above the mouth of the test tube turns blue.

Watch out!

$NH_4NO_2(aq)$, made by mixing $NH_4Cl(aq)$ and $NaNO_2(aq)$, could decompose explosively when heated!

Group VI:
oxygen and sulphur

Greek: oxys – *acid*
gennao – *I produce*

Antoine Laurent Lavoisier

Action point

Make a table for group VI similar to the tables for groups IV and V: see pages 94 and 98.

Watch out!

$CO_2(g)$ suffocates and $CS_2(l)$ poisons as well as burning to $CO_2(g)$ and $SO_2(g)$ which poisons!

Checkpoint

Give three reactions in which oxygen is formed by the thermal decomposition of an inorganic compound.

The jargon

Photosynthesis is a plant process using sunlight and chlorophyll to convert CO_2 and H_2O into carbohydrates, $C_x(H_2O)_y$. *Trioxygen = ozone* – a very reactive allotrope of oxygen.

Only these first two non-metals are important for A-level. The trends with increasing atomic number down the group follow similar patterns to those of other p-block groups.

The elements

→ Both elements exist as simple covalent molecules, O_2 and S_8.
→ Both elements have allotropes: O_2 and O_3 (ozone) in the gas phase, rhombic and monoclinic sulphur in the solid (crystalline) state.
→ Both elements are components of rocks and minerals in the earth's crust and exist uncombined (oxygen in air, sulphur in the ground).
→ Oxygen is extracted (blue liquid, b.p. $-183\,°C$) from the atmosphere by fractional distillation of liquid air while sulphur is extracted from the ground in the Frasch process.
→ Both elements combine with metals to form anions (O^{2-} and S^{2-}) and with non-metals to form covalent molecules (CO_2 and CS_2).

Oxygen

→ Oxygen is formed by thermal decomposition of metal nitrates and by electrolytic decomposition of water using an inert anode and a suitable electrolyte (e.g. H_2SO_4) to improve conductivity:

$$4OH^-(aq) \rightarrow 2H_2O(l) + O_2(g) + 4e^- \text{ (anodic oxidation } -2 \text{ to } 0)$$

→ Oxygen is produced by green plants during the photosynthesis of carbohydrates from atmospheric carbon dioxide and water:

$$6CO_2(g) + 6H_2O(l) \rightarrow C_6H_{12}O_6(s) + 6O_2(g)$$

Uses of oxygen

→ steel production – burns off unwanted carbon in pig-iron
→ metal cutting and welding – burns $C_2H_2(g)$ in oxyacetylene torch
→ medical uses and as ozone for air and water sterilization

Trioxygen, O_3 occurs in the stratosphere in the ozone layer which absorbs and protects the earth from harmful UV radiation.

→ Ozone is formed when an electric discharge passes through air (or oxygen): $3O_2(g) \rightarrow 2O_3(g) \rightarrow 3O_2(g)$.

This ozonized air (or oxygen) may contain up to 10% ozone. The O_3 is *not* in an equilibrium with O_2 but its decomposition back to oxygen is very slow at room temperature.

Sulphur

Rhombic (the more stable) and monoclinic sulphur crystals consist of closely packed S_8 covalent molecules (buckled rings).

On heating in a test tube, sulphur melts to a yellow liquid, darkens, and suddenly becomes unexpectedly extremely viscous. On heating further, the orange-red jelly becomes less viscous and turns almost black as the sulphur nears its boiling point. At the mouth of the tube a blue flame may appear.

→ Most metals including silver form sulphides many of which occur naturally as important ores mined for the extraction of the metal.
→ Sulphur burns in oxygen to sulphur dioxide.

Oxides and oxoacids

Binary compounds with oxygen exist for most elements. We classify these in several ways. At A-level you can classify them by *structure* (ionic, covalent, giant covalent, peroxide, superoxide) and by *acid–base behaviour* (basic, acidic, amphoteric and neutral).

Sulphur dioxide and trioxide

→ Sulphur dioxide liquefies easily, dissolves well in water forming a weak acid and forms sulphites with alkalis:

$$SO_2(g) + H_2O(l) \rightarrow H_2SO_3(aq) \text{ (sulphurous acid)}$$

If you add dilute hydrochloric acid to a sulphite, you will detect $SO_2(g)$ given off. Aqueous sulphur dioxide and sulphites are reducing agents that turn $Cr_2O_7^{2-}(aq)/H^+(aq)$ from orange to green.

→ Sulphur trioxide has a trigonal planar molecule in the gaseous state and forms deliquescent crystalline needles and a polymer in the solid state.

→ Sulphur trioxide is formed from SO_2 and O_2 in the Contact process for the manufacture of sulphuric acid, H_2SO_4.

Sulphurous and sulphuric acid

→ Both acids are dibasic, forming normal and acid salts with alkalis.
→ $H_2SO_3(aq)$ is a weak acid and $H_2SO_4(aq)$ is a strong acid:

$$H_2SO_3(aq) \rightleftharpoons H^+(aq) + HSO_3^-(aq) \rightleftharpoons H^+(aq) + SO_3^{2-}(aq)$$
$$H_2SO_4(aq) \rightarrow H^+(aq) + HSO_4^-(aq) \rightleftharpoons H^+(aq) + SO_4^{2-}(aq)$$

→ Sulphurous acid exists only in aqueous solution but (anhydrous) sulphuric acid is a viscous colourless liquid (b.p. 280 °C).

To make dilute aqueous sulphuric acid we add drops of conc. H_2SO_4 to a large volume of water stirred vigorously to dissipate the large amount of heat produced. Conc. H_2SO_4 can be reduced to SO_2 by metals and non-metals, e.g. $Zn(s) + 2H_2SO_4(l) \rightarrow ZnSO_4(aq) + SO_2(g) + 2H_2O(l)$ and $C(s) + H_2SO_4(l) \rightarrow CO_2(g) + 2SO_2(g) + 2H_2O(l)$. Conc. H_2SO_4 can be used as a dehydrating agent: $(COOH)_2(s) \rightarrow CO(g) + CO_2(g)$.

→ The sulphuric acid molecule and the sulphate anion have tetrahedral structures, the SO_4^{2-} being delocalized with four identical bonds between S—O and S=O equal in length.

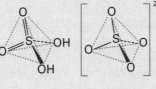

Action point

Make a table to show the two classes (ionic or covalent) of binary oxygen compounds and include Al_2O_3, CsO_2, CaO, CO_2, CO, K_2O_2, NO_2, SiO_2, Na_2O as examples.

The jargon

Sulphurous acid = sulphuric(IV) acid
Sulphuric acid = sulphuric(VI) acid
Sulphite = sulphate(IV)
Sulphate = sulphate(VI)

Watch out!

Sulphur dioxide may be used to sterilize food and wine but it is toxic. Fortunately its characteristic sharp smell will warn you of danger.

Links

See page 114: the Contact process.

The jargon

Normal salt = *all* protons donated:
e.g. $H_3PO_4 \rightarrow Na_3PO_4$
Acid salt = *not* all protons donated:
e.g. $H_3PO_4 \rightarrow Na_2HPO_4$ and NaH_2PO_4

Watch out!

pH of acid salt solutions need not be <7! $Na_2HPO_4(aq)$ is alkaline: pH > 7. Only the first ionization of $H_2SO_4(aq)$ is complete. The second is much weaker with pK_a = 1.99.

Exam question (10 min)   answer: page 131

(a) Suggest an explanation for (i) the changes observed when sulphur is heated and (ii) what happens when the hot liquid sulphur is poured quickly into cold water.

(b) Write equations for the reaction of sulphur with (i) magnesium and (ii) chlorine

(c) Suggest why the oxidation states of sulphur in Na_2S and SF_6 are different.

Group VI: water and hydrogen peroxide

Both oxygen and sulphur form hydrides, H_2O and H_2S. We die without water. We can be poisoned by hydrogen sulphide (TLV 20). Oxygen forms a peroxide, H_2O_2. Sulphur can form aqueous polysulphide anions.

Watch out!

$H_2(g)$ reacts with $O_2(g)$ explosively and irreversibly but with boiling sulphur only very slightly and reversibly.

Checkpoint 1

What is the conjugate base of (a) H_3O^+ and (b) OH^-? Suggest why water may be seen as amphoteric.

Links

See pages 19–20: hydrogen bonding.

Water ●●●

→ Water is the most abundant liquid on our planet.
→ Water is a poor conductor (but not an insulator) and its self-ionization defines neutrality for aqueous acid–base systems:

$$H_2O(l) + H_2O(l) \rightleftharpoons H_3O^+(aq) + OH^-(aq)$$

Compared to most materials or to covalent compounds of similar molar mass, water has some exceptional properties:

1 high melting point and boiling temperatures
2 high specific heat capacity and heats of fusion and vaporization
3 solid (ice) less dense than the liquid on which it floats
4 density of the liquid has a maximum value at 4 °C

You should be able to account for these properties in terms of the structure, shape and bonding of the polar water molecule and the intermolecular forces in the liquid and in the open structure of the solid.

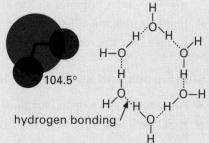

→ Water is an excellent polar solvent for ionic solids and many polar covalent compounds.

Water from limestone districts is 'hard' because it contains $Ca^{2+}(aq)$ which precipitates (soluble sodium) soaps as 'scum' (insoluble calcium soaps): $Ca^{2+}(aq) + 2C_{17}H_{35}COO^-(aq) \rightarrow (C_{17}H_{35}COO^-)_2Ca^{2+}(s)$. Some so-called 'temporary hardness' is removed by boiling if $HCO_3^-(aq)$ is present: $Ca^{2+}(aq) + 2HCO_3^-(aq) \rightarrow CaCO_3(s) + H_2O(l) + CO_2(g)$. The precipitated calcium carbonate is the limescale or kettle fur. $Mg^{2+}(aq)$ and any $Ca^{2+}(aq)$ not removed by boiling is 'permanent hardness'.

→ Synthetic ion-exchange resins (used in dishwashers) soften water by absorbing $Ca^{2+}(aq)$ and replacing it with $Na^+(aq)$.
→ Polyphosphates (included in washing powders) can soften water by forming complexes with the $Ca^{2+}(aq)$.

Many crystalline salts are hydrated, e.g. cobalt(II) chloride-6-water and copper(II) sulphate-5-water. Some or all of the water molecules are attached to the cation by dative covalent bonds. The O-atom in the H_2O donates the lone pair for sharing and forming the bond.

→ Water molecules can act as ligands to form complexes.
→ We use $CaCl_2(s)$ and $Na_2SO_4(s)$ as drying agents in organic chemistry because they readily form hydrates $CaCl_2 \cdot 6H_2O(s)$ and $Na_2SO_4 \cdot 10H_2O(s)$ but do not dissolve in organic liquids.
→ Water may act as a nucleophile in the hydrolysis of some inorganic and organic halogen compounds.

Checkpoint 2

Suggest why some dishwashers have a container that must be kept filled with granular salt (sodium chloride).

Watch out!

We prefer to use sodium sulphate as a drying agent because calcium chloride is deliquescent and may not be so easy to separate from the organic liquid.

Links

See pages 108–9: inorganic halides; and 149: acyl chlorides.

Water reacts as a reducing agent with fluorine because the halogen is a more powerful oxidant than oxygen itself and displaces it:

$$2F_2(g) + 2H_2O(l) \rightarrow 4HF(aq) + O_2(g)$$

→ Water normally reacts as an oxidant and is reduced to hydrogen by reaction with, for example, metals and methane:

$$2Na(s) + 2H_2O(l) \rightarrow 2NaOH(aq) + H_2(g)$$

$$CH_4(g) + 2H_2O(g) \xrightarrow[\text{nickel catalyst}]{900\,°C\ 30\ atm} CO(g) + 2H_2(g)$$

Hydrogen peroxide

H_2O_2 is formed in a variety of situations including glow discharge through water vapour, oxidation of hydrocarbons, exposing a hydrogen–oxygen mixture to intense UV light and during electrolysis of aqueous sulphuric acid and sulphates. In many of these cases, hydroxyl **free radicals** may be formed and, under suitable conditions, may combine to give the peroxide: $2HO\cdot \rightarrow HO{-}OH$.

→ Pure hydrogen peroxide is a pale blue liquid liable to decompose explosively and exothermically without warning into H_2O and O_2.

You should be able to use the VSEPR theory to predict a shape for the molecule.

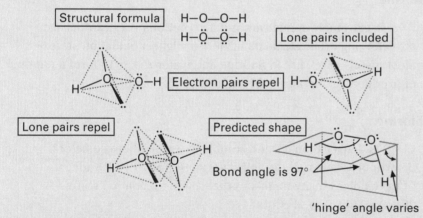

s-Block peroxides are ionic with O_2^{2-} ions in their structure
We could prepare aqueous hydrogen peroxide in the laboratory from barium peroxide and ice-cold aqueous sulphuric acid because barium sulphate is insoluble: $BaO_2(s) + H_2SO_4(aq) \rightarrow BaSO_4(s) + H_2O_2(aq)$. However, we usually buy aqueous hydrogen peroxide sold as '20 (or 100) volume solution'.

→ Peroxides can provide free radicals to initiate **polymerizations**.

Potassium manganate(VII) oxidizes hydrogen peroxide in excess $H_2SO_4(aq)$ quantitatively: $5H_2O_2(aq) + 2MnO_4^-(aq) + 6H^+(aq) \rightarrow 2Mn^{2+}(aq) + 8H_2O(l) + 5O_2(g)$.

Exam question (4 min)   answer: page 131

25.0 cm^3 of aqueous hydrogen peroxide in excess acid required 37.5 cm^3 of aqueous potassium manganate(VII), of concentration 0.018 7 mol dm^{-3}, for complete oxidation. Calculate the concentration of aqueous hydrogen peroxide in mol dm^{-3}.

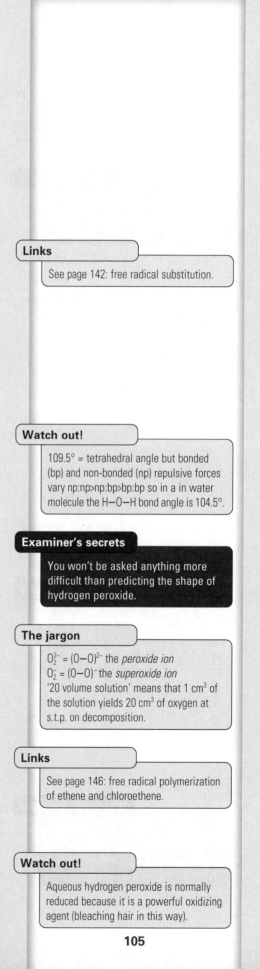

Links

See page 142: free radical substitution.

Watch out!

109.5° = tetrahedral angle but bonded (bp) and non-bonded (np) repulsive forces vary np:np>np:bp>bp:bp so in a in water molecule the H—O—H bond angle is 104.5°.

Examiner's secrets

You won't be asked anything more difficult than predicting the shape of hydrogen peroxide.

The jargon

O_2^{2-} = $(O{-}O)^{2-}$ the *peroxide ion*
O_2^- = $(O{-}O)^-$ the *superoxide ion*
'20 volume solution' means that 1 cm^3 of the solution yields 20 cm^3 of oxygen at s.t.p. on decomposition.

Links

See page 146: free radical polymerization of ethene and chloroethene.

Watch out!

Aqueous hydrogen peroxide is normally reduced because it is a powerful oxidizing agent (bleaching hair in this way).

Group VII: halogens and hydrogen halides

The halogens are very important at A-level. They are all non-metals and demonstrate very clear trends in their behaviour as you go down the group. Most syllabuses do not include fluorine but check yours in case.

The elements

→ All can have an oxidation number of –1 in covalent compounds and form a simple anion with a 1– charge.
→ Fluorine is a stronger oxidant than oxygen and forms compounds with all other elements except helium, neon and argon.
→ Unlike fluorine, the other halogens can have positive oxidation numbers and undergo disproportionation reactions.
→ All the halogens are toxic but chlorine is used to sterilize water and iodine has been used as an antiseptic agent.

Formula and state	Colour of element	Electronic configuration	M.p. /°C	B.p. /°C	$E(X–X)$ /kJ mol^{-1}	$E(H–X)$ /kJ mol^{-1}	Np
$F_2(g)$	pale yellow	$1s^22s^22p^5$	–220	–188	158	562	4.0
$Cl_2(g)$	greenish-yellow	$1s^22s^22p^63s^23p^5$	–101	–35	242	431	3.0
$Br_2(l)$	red-brown	$[Ar]3d^{10}4s^24p^5$	–8	59	193	366	2.8
$I_2(s)$	lustrous grey-black	$[Kr]4s^24p^64d^{10}5s^25p^5$	114	184	151	299	2.5

Fluorine

→ $F_2(g)$ is extremely dangerous and its reactions very exothermic, often forcing other elements into their highest oxidation state.
→ Most metals catch fire in fluorine and water reacts to form a number of products including O_2, O_3 and H_2O_2.

Chlorine

→ $Cl_2(g)$ is denser than air and soluble in water where some of it disproportionates: $Cl_2(aq) + H_2O(l) \rightleftharpoons HCl(aq) + HClO(aq)$.
→ $Cl_2(g)$ oxidizes many metals to their higher oxidation state: $2Fe(s) + 3Cl_2(g) \rightarrow 2FeCl_3(s)$.
→ $Cl_2(g)$ forms compounds with many non-metals (e.g. PCl_3 and PCl_5) but does not react directly with carbon, nitrogen and oxygen.
→ Chlorine can displace bromine and iodine from their compounds and from their aqueous ions in accord with the $E^{\ominus}$/V values for the following redox processes:

$$Cl_2(aq) + 2e^- \rightleftharpoons 2Cl^-(aq) \quad + 1.36$$
$$Br_2(aq) + 2e^- \rightleftharpoons 2Br^-(aq) \quad + 1.09$$
$$I_2(aq) + 2e^- \rightleftharpoons 2I^-(aq) \quad + 0.54$$
$$F_2(aq) + 2e^- \rightleftharpoons 2F^-(aq) \quad + 2.87$$

Note: Fluorine would never be used in aqueous displacement reactions. Chlorine disproportionates in NaOH(aq) to different oxidation states depending on the alkali concentration and temperature:

1   $Cl_2(g) + 2NaOH(aq) \rightarrow NaCl(aq) + NaClO(aq) + H_2O(l)$
2   $3Cl_2(g) + 6NaOH(aq) \rightarrow 5NaCl(aq) + NaClO_3(aq) + 3H_2O(l)$

Some manufacturing processes produce excess chlorine – a potential environmental health hazard. This chlorine can be absorbed by alkalis to produce for example NaClO (sodium chlorate(I) or sodium hypochlorite), the active ingredient in household bleach. $NaClO_3$ (sodium chlorate(V) or sodium chlorate) is a weed killer and oxidant.

Bromine and iodine

→ Bromine is moderately and iodine only slightly soluble in water.
→ Halogens dissolve in organic solvents, iodine solutions being brown in alcohols and violet in hydrocarbons or halogenoalkanes.
→ Br_2 and I_2 reactions are like those of chlorine but less vigorous.
→ $Br_2(aq)$ is rapidly decolourized by alkenes and $I_2(aq)$ forms a blue-black complex with starch (a very sensitive test for I_2 or starch).

Hydrogen halides

→ All four (volatile) hydrogen halides are released as gases from solid halides by hot (involatile) phosphoric acid: $Khal(s) + H_3PO_4(l) \rightarrow KH_2PO_4(s) + Hhal(g)$.
→ They dissolve readily in water to give (weak) hydrofluoric acid and strong hydrochloric, hydrobromic and hydroiodic acids.
→ Strong oxidizing agents like $KMnO_4(s)$ or $PbO_2(s)$ release the halogen when warmed with conc. HCl(aq), HBr(aq) and HI(aq): $PbO_2(s) + 4HCl(conc.aq) \rightarrow PbCl_2(s) + 2H_2O(l) + Cl_2(g)$.
→ We can detect hydrogen halides by the white smoke of ammonium halide formed exothermically on contact with ammonia gas.

Hydrogen fluoride

Hydrogen fluoride is prepared by the action of hot concentrated sulphuric acid on an ionic fluoride: $CaF_2(s) + H_2SO_4(l) \rightarrow CaSO_4(s) + 2HF(g)$. It etches glass.

→ Hydrogen fluoride is highly polar and hydrogen bonded.

High intermolecular forces are the reason why the boiling point (19.5 °C) is higher than that of HCl. Unusually strong hydrogen bonds cause dimers, H_2F_2, and lower the number of protons donated in water, so the aqueous acid is weak ($K_a = 7 \times 10^{-4}$ mol dm^{-3}). Sodium hydrogenfluoride, $NaHF_2(s)$, a salt with HF_2^- ions, exists.

→ Hydrofluoric acid is stored in polythene containers because it attacks and etches glass by reacting with the silicon dioxide: $SiO_2(s) + 6HF(aq) \rightarrow H_2SiF_6(aq)$.

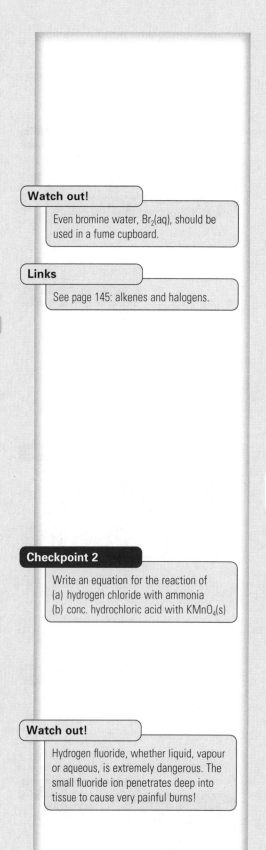

Watch out!

Even bromine water, $Br_2(aq)$, should be used in a fume cupboard.

Links

See page 145: alkenes and halogens.

Checkpoint 2

Write an equation for the reaction of
(a) hydrogen chloride with ammonia
(b) conc. hydrochloric acid with $KMnO_4(s)$

Watch out!

Hydrogen fluoride, whether liquid, vapour or aqueous, is extremely dangerous. The small fluoride ion penetrates deep into tissue to cause very painful burns!

Exam question (5 min)   answer: page 132

(a) Write an equation and give the oxidation number of the chlorine in each product when the gas reacts with (i) cold, dilute NaOH(aq), (ii) hot, concentrated NaOH(aq). (b) Predict the structure and shape of the HF_2^- ion.

Group VII: halides and interhalogen compounds

Now we look at the compounds formed between the halogens and hydrogen and their stability, which you can relate to the increasing bond length from HCl to HI. You will also see the ways in which solid and aqueous halides can be identified.

Links

See page 115: HCl manufacture.

Hydrogen chloride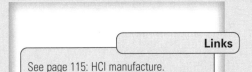

Hydrogen chloride is prepared by the action of hot concentrated sulphuric acid on an ionic chloride: $NaCl(s) + H_2SO_4(l) \rightarrow NaHSO_4(s) + HCl(g)$.

→ Hydrogen chloride is a colourless, choking gas that fumes in moist air and dissolves rapidly in water where as a strong acid it ionizes completely: $HCl(g) + H_2O(l) \rightarrow H_3O^+(aq) + Cl^-(aq)$.

Checkpoint 1

Deduce which of HBr and HI is the stronger reductant by finding the change in oxidation number of sulphur when each reduces sulphuric acid.

Hydrogen bromide and hydrogen iodide

Like hydrogen chloride, HBr(g) and HI(g) are colourless choking gases that fume in moist air and dissolve rapidly in water where as strong acids they ionize completely into hydrogen ions and halide ions.

→ Hydrogen bromide and hydrogen iodide are *not* prepared using hot conc. sulphuric acid because

1 they are reductants and would be oxidized by the acid:

$$2HBr(g) + H_2SO_4(l) \rightarrow 2H_2O(l) + SO_2(g) + Br_2(l) \text{ – brown fumes}$$
$$8HI(g) + H_2SO_4(l) \rightarrow 4H_2O(l) + H_2S(g) + 4I_2(s) \text{ – purple vapour}$$

2 they would thermally decompose:

$$2HBr(g) \rightleftharpoons H_2(g) + Br_2(g) \text{ and } 2HI(g) \rightleftharpoons H_2(g) + I_2(g)$$

→ Hydrogen bromide is formed when hydrogen burns in bromine vapour: $H_2(g) + Br_2(g) \rightarrow 2HBr(g)$.

It is not energetically feasible to prepare hydrogen iodide by direct combination of iodine with hydrogen – the reaction is endothermic. On the contrary, if you plunge a red hot silica rod into HI(g), you will see a purple vapour and a black precipitate (of iodine) form instantly.

Checkpoint 2

Write an equation for the reaction of sodium hydrogencarbonate with hydrochloric acid.

Hydrochloric acid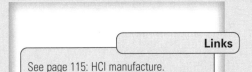

→ Hydrochloric acid is a solution of $H_3O^+(aq) + Cl^-(aq)$.

This non-oxidizing mineral acid is typical of the 'hydrohalidic' acids in their reactions as aqueous acids with alkalis, basic oxides, carbonates and hydrogencarbonates:

$$HCl(aq) + KOH(aq) \rightarrow KCl(aq) + H_2O(l)$$
$$2HCl(aq) + MgO(s) \rightarrow MgCl_2(aq) + H_2O(l)$$
$$2HCl(aq) + Na_2CO_3(s) \rightarrow 2NaCl(aq) + H_2O(l) + CO_2(g)$$

Halides

In organic compounds, halogen atoms are usually attached to carbon atoms by covalent bonds. Many inorganic halides are covalent or highly covalent in character. Binary s-block metal halides and the halides of d-block metals in their lower oxidation states are usually ionic or highly ionic in character.

- → Identifying a halide means converting it to hydrogen halide, $HX(g)$, or aqueous halide ion, $X^-(aq)$, and testing it.
- → You should know how to identify chloride, bromide and iodide in theory *and* practice but a fluoride in theory only.

Identifying solid halides

- → *General procedure*: warm a small sample with conc. $H_2SO_4(l)$
- → *General observation*: steamy fumes give white smoke with $NH_3(g)$

Halide	Specific observations
fluoride	fumes etch glass not protected by candle wax
chloride	no colour
bromide	brown colour
iodide	purple colour and grey-black solid

- → *Additional test*: mix halide with manganese(IV) oxide before adding conc. $H_2SO_4(l)$ and identify $Cl_2(g)$ with moist blue litmus
- → *Observation*: litmus paper turns from blue → red → white

Identifying aqueous halide ions

- → *General procedure*: acidify with $HNO_3(aq)$, add aqueous silver nitrate, $AgNO_3(aq)$, then make alkaline with aqueous ammonia
- → *General observation*: precipitate forms and then disappears

Halide	Specific observations
fluoride	no precipitate forms
chloride	white ppt complete 'dissolves' in ammonia
bromide	off-white ppt partly 'dissolves' in ammonia
iodide	pale yellow ppt not affected by ammonia

- → *Additional test*: add aqueous calcium chloride
- → *Observation*: white precipitate (of calcium fluoride)

Interhalogen compounds ○○○

Halogens combine with one another to form covalent compounds. All six possible diatomic combinations (ClF, BrF, IF, BrCl, ICl, IBr) exist and are named as the halide of the less reactive halogen.

Checkpoint 3

Suggest why
(a) we must (i) acidify the solution and (ii) use nitric acid to do so
(b) AgF is soluble in water but AgCl, AgBr and AgI are insoluble and increasingly so from AgCl to AgI
(c) calcium fluoride is insoluble in water even though most ionic fluorides are soluble

Watch out!

You may add drops of $Cl_2(aq)$ to $NaBr(aq)$, add an organic solvent and *not* see a red-brown colour because the displaced $Br_2(aq)$ reacts with the $Cl_2(aq)$ to form BrCl(aq)!

Checkpoint 4

What is the name of (a) ICl, (b) ICl_3?

Exam question (5 min) answer: page 132

(a) Write an equation for the reaction of (i) chlorine with aluminium and (ii) aqueous iodide ions with manganate(VII) ions in excess acid.

(b) Use the equation $IO_3^-(aq) + 5I^-(aq) + 6H^+(aq) \rightarrow 3I_2(aq) + 3H_2O(l)$ to calculate the concentration of 20 cm³ $KIO_3(aq)$ that liberates 0.15 mol of aqueous iodine molecules from excess potassium iodide solution.

Industrial chemistry: aluminium and carbon

Aluminium, oxygen and silicon are major constituents of many clays and rocks composing the earth's crust. Carbon is the major component of living things. Nitrogen and phosphorus are essential for plant growth and major constituents of fertilizers. Sulphuric acid is such an important industrial chemical that its annual production has been taken as an indication of a nation's economic prosperity.

Group III

Aluminium

Aluminium is an important 20th century metal. Every year worldwide 22 million tonnes are produced and a further 7 million tonnes are obtained by recycling. An American (Charles Hall) and a Frenchman (Paul Héroult) independently invented in the 19th century the method we still use today to extract the metal from its main source – the mineral *bauxite*, a hydrated oxide, $Al_2O_3 \cdot 2H_2O$.

Extraction of aluminium

Bauxite is mined mainly in Africa, Australia, South America and the West Indies.

Basic Fe_2O_3 and TiO_2 impurities in bauxite are removed from the *amphoteric* Al_2O_3 using sodium hydroxide at 150 °C and 4 atm to form a solution of sodium hydroxyaluminate and insoluble impurities:

$$Al_2O_3/Fe_2O_3/TiO_2(s) + 2NaOH(aq) + 3H_2O(l)$$
$$\rightarrow 2NaAl(OH)_4(aq) + Fe_2O_3/TiO_2(s)$$

After the impurities are removed, the solution is cooled and seeded with pure alumina to precipitate alumina trihydrate which decomposes on heating to 1 000 °C to give pure aluminium oxide.

Carbon forms oxides with liberated oxygen so graphite anodes must be renewed

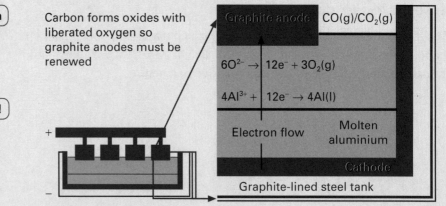

$$6O^{2-} \rightarrow 12e^- + 3O_2(g)$$

$$4Al^{3+} + 12e^- \rightarrow 4Al(l)$$

Graphite anode   CO(g)/CO₂(g)

Electron flow   Molten aluminium

Cathode

Graphite-lined steel tank

→ Alumina is dissolved in molten cryolite, $(Na^+)_3AlF_6^{3-}$, containing some calcium fluoride to lower the melting point to about 950 °C.
→ Replacement graphite anodes are usually made at the plant.
→ Small amounts of cryolite must be added occasionally.

Aluminium can be anodized and alloyed with other elements to produce low-density materials for specialist uses.

The jargon

A *mineral* is any substance that can be mined or that occurs naturally as a product of inorganic processes.

Action point

What important factors must be considered when choosing where to build a new industrial aluminium extraction and processing plant?

The jargon

Cryolite is a mineral ore found in Greenland. But Na_3AlF_6 used in the Hall–Héroult process is now synthetic.

Watch out!

Potentially it is a very reactive metal but it has a tough oxide coating that protects the surface from attack.

Aluminium compounds

Aluminium oxide: major industrial uses are in glass, paper and pottery manufacture, in water treatment, as a refractory, an abrasive and insulator, and as a catalyst promoter and support in the chemical and petroleum industries.

→ Anodizing is an electrolytic process to thicken the aluminium oxide layer so it adsorbs dyes and permanently colours the metal surface.

Aluminium sulphate: very effective coagulating agent (Al^{3+}(aq) ions) used for water treatment but suspected of causing brain damage.

Aluminium triethyl, $Al(C_2H_5)_3$: used with titanium(IV) chloride in Ziegler–Natta catalysts to produce high-density poly(ethene) and poly(propene).

Lithium aluminium hydride, $LiAlH_4$: a versatile organic reagent used in dry ether to reduce, for example, an unsaturated carboxylic acid to a primary alcohol without affecting the carbon–carbon double bond.

Group IV ●●●

Carbon

→ Coke is an essential raw material in the iron and steel industry.
→ Diamonds are used in industrial drills and cutting tools.
→ Graphite is used as a lubricant, as electrodes in industrial electrolytic extraction processes and in pencils.
→ Carbon fibres are used to produce lightweight but very strong materials incorporated into racing cars, tennis rackets, etc.

Carbon compounds

Carbon monoxide: dangerous pollutant (from incomplete combustion of petrol and other hydrocarbons), reducing iron oxides to iron in the blast furnace, combines with hydrogen (when heated under pressure with ZnO/Cr_2O_3 catalysts) in the industrial production of methanol.

Carbon dioxide: acidic gas, produced by combustion of carbon-containing fuels, converted (with water) photosynthetically by plants into carbohydrates, increases now causing the 'greenhouse effect'.

Carbonates: sodium carbonate (Na_2CO_3) and calcium carbonate ($CaCO_3$) are important industrial chemicals.

Hydrocarbons: natural gas (methane), petroleum, etc., used as fuels and feedstock in the petrochemical industry.

Interstitial carbides: non-stoichiometric compounds of carbon with metals like tungsten carbide, extremely hard and used in drill bits.

Action point

Find out what happened when a lorry load of aluminium sulphate was dumped into the wrong tank at the Camelford water works.

The jargon

The systematic name for $LiAlH_4$ is *lithium tetrahydridoaluminate(III)*.

Checkpoint

Suggest why $LiAlH_4$ must be used in dry ether.

Links

See pages 162–3: petrochemical industry.

Exam question (5 min)   answer: page 132

(a) For the manufacture of aluminium, suggest (i) how the Fe_2O_3 and TiO_2 impurities would be removed from the hot aqueous sodium hydroxyaluminate, (ii) what would happen if the bauxite were not purified before being electrolysed and (iii) why the cryolite must be replenished.

(b) State how and explain why aluminium is anodized.

Industrial chemistry: silicon and nitrogen

The jargon

Si = silicon (element)
SiO_2 = silica (silicon dioxide)

Checkpoint 1

What is the oxidation number of Si in its compounds?

The jargon

Zone melting = zone refining
In a *semiconductor* conductivity rises with rise in temperature (*n*- or *p*-type semiconductors).

The formation of ammonia in the Haber process and its subsequent oxidation are very important chemical processes. Ammonia is needed for fertilizer production, which uses 80% of all NH_3 made, and the production of nitric acid, explosives and nylon.

Silicon

The element is extracted by heating quartz in a carbon arc:

$$SiO_2 + 2C \rightarrow Si + 2CO$$

Very pure silicon is produced by

1. making silicon chloride: $Si(s) + 2Cl_2(g) \rightarrow SiCl_4(l)$
2. purifying the chloride by fractional distillation
3. reducing the chloride using very pure magnesium:
    $SiCl_4(l) + 2Mg(s) \rightarrow Si(s) + 2MgCl_2(s)$
4. purifying the element by zone melting

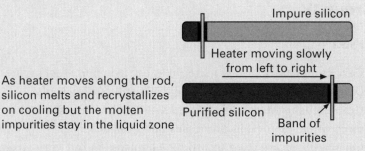

As heater moves along the rod, silicon melts and recrystallizes on cooling but the molten impurities stay in the liquid zone

Impure silicon
Heater moving slowly from left to right
Purified silicon
Band of impurities

The production of extremely pure semiconducting silicon crystals led to silicon chips, modern electronics and computer technology.

Silicates

→ Glass is a transparent mixture of silicates manufactured by melting sandstone, limestone and soda in a high-temperature furnace.

The complex reactions may be represented by two simplified equations:

$$CaCO_3(s) + SiO_2(s) \rightarrow CaSiO_3(l) + CO_2(g)$$
$$Na_2CO_3(s) + SiO_2(s) \rightarrow Na_2SiO_3(l) + CO_2(g)$$

The jargon

Clay is a complicated mixture of silicates and aluminosilicates.
Portland cement is so-called because its inventor thought it looked like the stone from the quarries in Portland, Dorset.

→ Cement is a complex mixture manufactured by heating limestone and clay in a rotary kiln, grinding and then mixing with gypsum.

The complex reactions may be represented by a simplified equation:

$$CaO \cdot Al_2O_3 \cdot 2SiO_2(s) + 8CaCO_3(s) \rightarrow 2Ca_3SiO_5(s) + Ca_3Al_2O_6(s) + 8CO_2(g)$$

The setting of cement is a complex exothermic process involving water stirred into the powder and carbon dioxide absorbed from the air.

→ Asbestos is an industrially useful heat-resistant fibrous mineral.

Checkpoint 2

Why is the use of blue asbestos prohibited and the use of white asbestos carefully controlled?

Group V

→ Nitrogen is produced by fractional distillation of liquid air.
→ Nitrogen is the raw material for the large-scale industrial production of ammonia and nitric acid.

Manufacture of ammonia by the Haber process

→ Nitrogen (with argon) comes from the air.
→ Hydrogen may come from water gas, steam reforming of natural gas, or catalytic reforming processes in the petroleum industry:

$$N_2(g) + 3H_2(g) \xrightarrow[\text{promoted iron catalyst}]{450 \,°C \ 200 \ \text{atm}} 2NH_3(g); \Delta H = -92 \text{ kJ mol}^{-1}$$

→ The heterogeneous catalyst is made by putting a mixture of oxides of iron, potassium, calcium and aluminium into the reaction vessel.

According to the **law of chemical equilibrium** formation of ammonia is favoured by high pressure and low temperature. According to the **kinetic molecular theory** rate of formation is favoured by high temperature and catalysts. In practice,

→ 15% conversion is achieved under optimum conditions
→ gases from the reaction vessel are cooled while still under pressure
→ liquid ammonia is removed as it forms

The major industrial use of ammonia is in the manufacture of fertilizers, nylon and nitric acid.

Manufacture of nitric acid by the Ostwald process

$$4NH_3(g) + 5O_2(g) \underset{\text{at 900 °C}}{\overset{\text{5 atm pressure} \quad \text{Pt/Rh gauze}}{\rightleftharpoons}} 4NO(g) + 6H_2O(g); \Delta H = -994 \text{ kJ}$$

keeps gauze catalyst hot

The gaseous product is compressed, cooled and mixed with air so the nitrogen oxide can spontaneously oxidize to nitrogen dioxide and hydrate to nitric acid containing around 60% HNO_3 by mass:

$$2NO(g) + O_2(g) \rightarrow 2NO_2(g)$$
$$4NO_2(g) + O_2(g) + 2H_2O(l) \rightarrow 4HNO_3(aq)$$

The major industrial use of nitric acid is the production of ammonium nitrate (as an explosive in quarrying and a fertilizer in agriculture) and in the **nitration** of organic compounds to produce explosives such as nitroglycerine, nitrocellulose and trinitrotoluene.

The jargon

Natural gas = methane
Water gas = CO(g) + H₂(g) formed by passing steam over white-hot coke
Catalytic reforming: see page 162
Steam reforming = CH₄(g) + H₂O(g) under pressure over heated Ni catalyst to form CO(g) + 3H₂(g)
Promoted iron catalyst = finely divided iron mixed with oxides of K, Ca and Al to enhance its activity

Checkpoint 3

What happens to the 85% unconverted hydrogen and nitrogen?

Exam question (10 min)   answer: page 132

The Haber process *fixes nitrogen* by synthesizing ammonia from nitrogen and hydrogen *under optimum conditions* using iron as a *heterogeneous catalyst*.

(a) Explain what is meant by the terms in italics.
(b) For the synthesis of ammonia, write an expression for K_p and use it to explain (i) why the Haber process operates at high pressure and (ii) why the gases are cooled while still under pressure and liquid ammonia is removed continuously as it forms from the reaction vessel.
(c) (i) Suggest why the catalyst is made in the reaction vessel by reducing iron(III) oxide and (ii) write a balanced equation for the reduction.

113

Industrial chemistry: sulphur and the halogens

Sulphur is extracted from the ground by the Frasch process or by desulphurization of fossil fuels, and used to produce sulphuric acid. Sodium chloride is mined or extracted from salt deposits as brine and electrolyzed to produce chlorine and sodium hydroxide.

Checkpoint 1

(a) Suggest what would happen if the gases were not (i) purified, (ii) cooled in heat exchangers.
(b) What is meant by (i) heterogeneous catalyst, (ii) promoter?

The Contact process for sulphuric acid ●●●

→ Sulphur dioxide comes from burning sulphur in air or from roasting sulphide minerals in some metal extraction plants:

$$S(s) + O_2(s) \rightarrow SO_2(g); \quad ZnS(s) + 1\tfrac{1}{2}O_2(s) \rightarrow ZnO(s) + SO_2(g)$$

→ Sulphur dioxide and excess air are purified and passed through a series of alternate reaction vessels and heat exchangers:

$$2SO_2(g) + O_2(g) \xrightleftharpoons[\text{at 430 °C}]{V_2O_5 \text{ catalyst}} 2SO_3(g); \quad \Delta H = -197 \text{ kJ}$$

removed by heat exchangers

The jargon

Mist is liquid droplets suspended in air. *Oleum* (fuming sulphuric acid) is a 'solution' of $SO_3(s)$ in $H_2SO_4(l)$.

→ The reaction vessels hold a heterogeneous catalyst of vanadium(V) oxide promoted by potassium sulphate and supported on silica.
→ A heat exchanger cools the gases to about 430 °C before they enter a reaction vessel to form sulphur trioxide exothermically.

In modern plants, conversion of SO_2 to SO_3 is almost complete.

→ Sulphur trioxide gas is absorbed in 98% sulphuric acid to form oleum ($H_2S_2O_7$) as the final product or for dilution to $H_2SO_4(l)$.
→ Sulphur trioxide gas is *not* added to water alone because that would produce an inconvenient stable acid mist.

Examiner's secrets

This is a favourite topic in exams. You will be asked to explain the chemical principles behind the Contact process.

Any unreacted sulphur dioxide is passed through an extra converter as an environmental control to minimize SO_2 emissions.

The optimum operating conditions for the process follow from the law of chemical equilibrium and the principles of reaction kinetics.

Links

See page 55: law of chemical equilibrium.

$$\frac{p_{SO_3}^2}{p_{SO_2}^2 \times p_{O_2}} = K_p$$

Checkpoint 2

Explain how the law of chemical equilibrium predicts that the yield of SO_3 in this reversible reaction would improve at high pressure and low temperature.

High pressure would improve the equilibrium yield of sulphur trioxide and increase its rate of formation. In practice the slight improvement would not justify the expense of high-pressure plant.

Low temperature would improve the equilibrium yield of sulphur trioxide but decrease its rate of formation.

Heterogeneous catalyst would not improve the equilibrium yield of sulphur trioxide but would increase its rate of formation.

Optimum conditions

→ enough pressure (about 2 atm) to pump gases through the reaction vessels and heat exchangers

→ compromise temperature to achieve a reasonable yield at a reasonable rate

→ solid catalyst to speed up the rate of attainment of equilibrium

Group VII ●●●

Manufacture of chlorine

Chlorine is produced by the electrolysis of chlorides during the manufacture of sodium hydroxide and reactive metals. Two important electrolytic processes may be summarized as follows:

$$NaCl(aq) + H_2O(l) \xrightarrow{\text{D.C. electricity}} \underset{\text{at the cathode}}{NaOH(aq) + H_2(g)} + \underset{\text{at the anode}}{Cl_2(g)}$$

Links

See page 90: sodium hydroxide and chlorine manufacture.

Uses of chlorine

→ water purification

→ manufacture of PVC, organic chemicals and chlorinated solvents

→ manufacture of hydrochloric acid and inorganic chemicals such as $NaClO(aq)$ and $NaClO_3(s)$

Checkpoint 3

What are CFCs and why are they banned in the EU?

Manufacture of bromine and iodine

→ Bromine is mostly extracted from acidified sea water by oxidizing bromide ions with chlorine in a non-metal displacement reaction:

$$Cl_2(g) + 2Br^-(aq) \rightarrow 2Cl^-(aq) + Br_2(g)$$

→ Iodine is mostly extracted by reducing sodium iodate(V) from Chile saltpetre (sodium nitrate) deposits using sulphur dioxide:

$$IO_3^-(aq) + 3SO_2(g) + 3H_2O(l) \rightarrow I^-(aq) + 3H_2SO_4(aq)$$
$$IO_3^-(aq) + 5I^-(aq) + 6H^+(aq) \rightarrow 3I_2(s) + 3H_2O(l)$$

Hydrogen chloride

Hydrogen chloride is manufactured by burning hydrogen in chlorine, both gases coming from the electrolysis of $NaCl(aq)$.

Exam question (20 min)   answer: page 133

(a) Outline how sulphur dioxide is converted to sulphuric acid on an industrial scale.

(b) Describe and explain what is observed on warming in a test tube a few drops of concentrated sulphuric acid added to (i) NaCl(s), (ii) KNO_3(s), (iii) KI(s), (iv) a mixture of ethanoic acid and pentan-1-ol.

(c) State the main industrial source of (i) bromine and (ii) iodine.

(d) Explain in outline how each halogen in (c) is obtained on a large scale.

(e) Explain briefly how and why aqueous silver nitrate can be used to distinguish aqueous chloride, bromide and iodide but not fluoride ions.

(f) State *two* chemicals manufactured from chlorine on the industrial scale and state one important commercial use for each chemical.

Group VIII (or 0): the noble gases

In 1785 Henry Cavendish, an eccentric genius, suspected that less than 1 in 120 parts of our atmosphere is unreactive and different from the rest of the air. More than 100 years later, Rayleigh and Ramsay discovered that air contained about 0.94% of a gas that was more inert than nitrogen itself.

The elements ●●●

When these elements were first added to the periodic table they were called the *inert gases* and labelled *group 0*.

	Electronic configuration	E_{m1}/kJ mol^{-1}	M.p./°C	B.p./°C
He	$1s^2$	2 372	−272	−269
Ne	$1s^2 2s^2 2p^6$	2 080	−249	−246
Ar	$1s^2 2s^2 2p^6 3s^2 3p^6$	1 519	−189	−186
Kr	$1s^2 2s^2 2p^6 3s^2 3p^6 3d^{10} 4s^2 4p^6$	13 511	−157	−152
Xe	$[Kr]4d^{10}5s^2 5p^6$	1 170	−112	−108
Rn	$[Xe]4f^{14}5d^{10}6s^2 6p^6$	1 037	−71	−62

Helium

Janssen found evidence of an undiscovered element (helium) during the solar eclipse in 1868 when he observed a new line in the sun's spectrum. In 1895 Ramsay discovered helium in *clevite*, a uranium mineral. Two Swedish chemists, Cleve and Langlet, also made the same discovery independently of Ramsay.

→ Helium is obtained by the fractional distillation of liquid air.
→ Most commercial helium comes from natural gas and petroleum.

Helium in the earth's crust is from α decay of heavy nuclides.

→ An α particle is a nucleus of two protons and two neutrons.
→ A helium atom forms when an α particle gains two electrons.

Argon (and neon, krypton and xenon)

→ Argon is the most abundant noble gas ($\approx$ 1%) in the atmosphere.
→ Argon (neon, krypton and xenon) are separated from the atmosphere by fractional distillation of liquid air.

Radon

→ Radon is a radioactive alpha emitter formed in the radioactive decay of heavy nuclides such as ^{238}U.

Over thousands of years radon has accumulated in certain rocks. In some places the gas has been detected inside houses and its presence has been linked to the incidence of some types of cancers.

→ The radon is thought to seep into houses from the rocks and to be a health risk.
→ One remedy is increased interior and under-floor ventilation to minimize the concentration of radon gas inside the house.

The jargon

Greek:
Helios – the sun
Neos – new
Argos – idle one
Kryptos – hidden
Xenos – guest, stranger or foreigner

Checkpoint 1

What are the noble metals?

Uses of the noble gases   ●●●

Helium

→ weather balloons and airships
→ underwater breathing apparatus (80% helium and 20% oxygen)
→ low temperature physics and low temperature research

Argon

→ largest use as an *inert atmosphere* in arc cutting and welding, in titanium production, in silicon and germanium zone refining, and in electric light bulbs to inhibit evaporation of the filaments

Neon, krypton and xenon

→ different low pressure mixtures of these three gases are used to fill fluorescent tubes and produce different coloured lights

The chemistry of the noble gases   ●●●

Until 1962 chemists had not made any true noble gas compound although they thought it should be possible to make $XePtF_6$ in view of the similarity of the first ionization energy of the oxygen *molecule* ($E_{m1} = 1175 \text{ kJ mol}^{-1}$) and xenon *atom* ($E_{m1} = 1170 \text{ kJ mol}^{-1}$) and the existence of $O_2^+PtF_6^-$.

→ In 1962, Neil Bartlett reported the preparation of $XePtF_6$.

Within a few months of Bartlett's announcement, chemists had prepared xenon fluorides by direct combination of the two elements.

→ Xenon and fluorine combine at 6 atm pressure when passed through a nickel tube heated to 400 °C:

$$Xe(g) + F_2(g) \rightarrow XeF_2(s)$$
$$Xe(g) + 2F_2(g) \rightarrow XeF_4(s)$$
$$Xe(g) + 3F_2(g) \rightarrow XeF_6(s)$$

The three xenon fluorides are all colourless crystals. We can store them indefinitely in nickel vessels but the higher fluorides are hydrolysed by a trace of water. The fluorides hydrolyse to xenon oxofluorides and xenon oxides.

The majority of noble gas compounds are the compounds of xenon. Krypton difluoride and radon difluoride have been prepared. Ever since 1962 when Bartlett made a compound of Xenon, we have called Group VIII in the periodic table the noble gases.

Checkpoint 2

What is meant by the molar first ionization energy of the oxygen molecule?

Watch out!

$E_{m1} = 1314 \text{ kJ mol}^{-1}$ for atomic oxygen.

Links

See page 118: d-block catalysts.

Exam question (8 min)   answer: page 133

(a) Write the ground state electronic configuration, in terms of s and p electrons, for an isolated argon atom and use it to explain the stability of chloride and potassium ions relative to chlorine gas and potassium metal.

(b) State how XeF_2 and XeF_4 may be prepared and deduce the shapes of the molecules.

1st transition series: metals, aqueous ions and redox

The colour and variety of oxidation states of the transition elements make them a fascinating area of study but quite tricky at times!

The jargon

Sc–Zn = first series of ten d-block elements with inner d-orbitals filling
Ti–Cu = eight metals with incomplete d-orbitals in some compounds
Sc, Zn = two metals with no incomplete d-orbitals in any compound.

The elements

Physical properties and uses of the metals

Unlike the physical properties of elements in a row across the p-block, the typical metallic properties of these d-block elements show little or no change with increasing atomic number from scandium to zinc.

→ They are all ductile and malleable, lustrous and sonorous hard metals with high m.p.s, b.p.s, tensile strength and electrical conductivity.

→ They readily form *alloys* and *interstitial compounds* because they have similar properties and atom size.

You should tie the above properties of d-block elements to some of their main uses, e.g. electrical and thermal conductivity for electrical wire (copper) and heating elements (nichrome), lustre and hardness for coins and chrome plating, sonority for musical instruments, etc.

→ The key to the characteristic physical and chemical properties of transition metals (Ti–Cu) is their ground state electronic configurations and *incomplete 3d-subshells*

Watch out!

Only one 4s electron for Cr and Cu.

Examiner's secrets

You should be able to explain that the ground state configuration for Cr and Cu is $4s^1$ (not $4s^2$) because five half-full d-orbitals and five full d-orbitals are particularly stable arrangements.

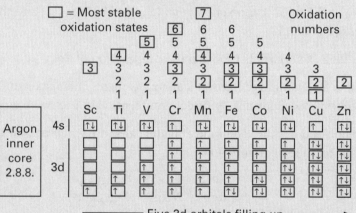

→ Transition elements are efficient *heterogeneous catalysts*.

For example, iron (in the Haber synthesis of ammonia) and nickel (in the hydrogenation of unsaturated oils) can use their d-orbitals when adsorbing the reacting gases to lower the activation energy barrier.

Links

See page 11: electronic configurations.

Links

See page 48: catalysis.

Extraction of the transition metals

→ The metals are usually obtained by chemical or electrolytic reduction of their mineral ores, e.g. iron in the blast furnace:

$$Fe_2O_3(s) + 3CO(g) \rightleftharpoons 2Fe(l) + 3CO_2(g)$$

The metals are purified by various methods.

→ Copper is refined electrolytically and titanium by a process involving decomposition of TiI_4 on a hot wire.

Checkpoint 1

Write an equation for (a) sodium reducing titanium(IV) chloride to titanium, (b) the extraction and refining of nickel using carbon monoxide, and (c) the ion–electron half-reaction for the reduction of copper(II) ions at the cathode during electrolytic refining.

Chemical properties

→ The metals form a variety of compounds with non-metals and react to varying extents with water and acids: e.g.

$$3Fe(s) + 4H_2O(g) \rightarrow Fe_3O_4(s) + 4H_2(g)$$
$$Fe(s) + 2HCl(g) \rightarrow FeCl_2(s) + H_2(g)$$
$$2Fe(s) + 3Cl_2(g) \rightarrow 2FeCl_3(s)$$

→ Rusting (and its prevention by a sacrificial metal) involves electrochemical reactions that include oxidation of iron (and of the sacrificial metal) and reduction of oxygen: e.g.

$$Fe(s) \rightarrow Fe^{2+}(aq) + 2e^-, \text{ then } Fe^{2+}(aq) \rightarrow Fe^{3+}(aq) + e^-$$
$$Zn(s) \rightarrow Zn^{2+}(aq) + 2e^-, \text{ then } O_2(g) + 2H_2O(l) + 4e^- \rightarrow 4OH^-(aq)$$

The transitional metal ions

The characteristic features of transition metal compounds are

→ variety and relative stability of oxidation states
→ formation and stability of various complexes
→ ability to act as catalysts

These features depend on d-orbitals being used for bonding.

For A-level the emphasis is on the *aqueous metal ions*. Most aqueous solutions of d-block compounds have distinctive colours associated with transition metal cations and/or oxoanions. You probably already associate blue with copper(II) sulphate, Cu^{2+}(aq) and purple with potassium manganate(VII), MnO_4^-(aq).

→ Colours of aqueous solutions change when transition metal ions take part in electron-transfer or ligand-transfer reactions.

We explain the colours of these aqueous solutions by electrons in the transition metal ions changing d-orbitals. The associated energy changes result in light (energy) of different colours being absorbed.

Redox reactions

Now is a good time to revise oxidation number rules, ion–electron half-equations, electrochemical cells and standard electrode potentials. We interpret a redox reaction of aqueous transition metal ions as an electron transfer and use $E^{\ominus}$ values to predict its feasibility.

For the elements Sc to Zn:

→ Stability of the higher oxidation states decreases from Sc to Zn.
→ Stability of the +2 oxidation state relative to the +3 oxidation state increases from Sc to Zn.
→ Greater stability of Mn^{2+} with respect to Mn^{3+} and Fe^{3+} with respect to Fe^{2+} may be related to the stability of the half-filled d-subshell.
→ Ions tend to exist as aqua cations in low oxidation states and oxoanions in higher oxidation states.

Watch out!

In these reactions the gases are passed over heated iron which must be kept red-hot except in the third case when the metal 'burns' in chlorine.

Examiner's secret

Favourite questions are about redox reactions, complex ion formation and homogeneous catalysis involving aqueous transition metal ions. The top four elements are Mn, Cr, Fe and Cu.

Links

See page 158: UV and visible spectroscopy.

Checkpoint 2

Write the formulae of (a) the chlorides and (b) the sulphates of the first row of d-block elements in their most stable oxidation states.

Exam question (8 min)   answer: page 133

(a) Explain briefly why zinc may be regarded as a d-block element but not a transition metal.

(b) State and account for the colour changes in solution when an excess of granulated zinc is added to aqueous sulphuric acid containing
(i) $K_2Cr_2O_7$(aq) and (ii) NH_4VO_3(aq).

1st transition series: redox reactions and

Transition metal ions undergo a variety of redox reactions, form a range of complexes and catalyze many reactions.

The jargon

Titrimetric analysis =

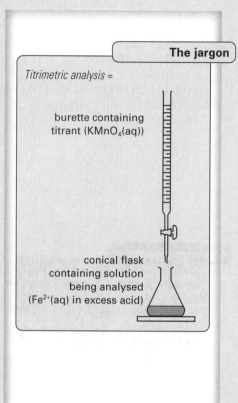

burette containing
titrant (KMnO$_4$(aq))

conical flask
containing solution
being analysed
(Fe^{2+}(aq) in excess acid)

Checkpoint

Calculate the standard free energy change for the redox reaction using the value +0.74 V and the expression
$$\Delta G^{\ominus} = -nFE^{\ominus}$$

Links

See pages 48–9: homogeneous catalysis.

Redox titrations ●●●

Potassium manganate(VII) and potassium dichromate(VI) are important oxidizing agents in titrimetric analysis. When they are used with excess sulphuric acid the reactions are quantitative and their reduction is represented by the following ion–electron half-equations:

$$MnO_4^-(aq) + 8H^+(aq) + 5e^- \rightarrow Mn^{2+}(aq) + 4H_2O(l)$$
$$Cr_2O_7^{2-}(aq) + 14H^+(aq) + 6e^- \rightarrow Cr^{3+}(aq) + 7H_2O(l)$$

When you run aqueous potassium manganate(VII) solution into aqueous iron(II) sulphate in excess sulphuric acid, the purple colour disappears as the iron(II) ions are oxidized to iron(III) ions:

$$Fe^{2+}(aq) \rightarrow Fe^{3+}(aq) + e^-$$

The oxidation number of manganese falls from +7 to +2 (5 units up) and that of iron rises from +2 to +3 (1 unit down).

1 mol MnO$_4^-$(aq) must oxidize 5 mol Fe^{2+}(aq) to balance the e$^-$s

We can combine the reduction and oxidation half-equations to eliminate the electrons and get the stoichiometric redox equation:

$$MnO_4^-(aq) + 8H^+(aq) + 5Fe^{2+}(aq) \rightarrow Mn^{2+}(aq) + 4H_2O(l) + 5Fe^{3+}(aq)$$

We could use this reaction in an electrochemical cell:

$$Pt\,|[Fe^{2+}(aq),Fe^{3+}(aq)]\vdots\vdots[MnO_4^-(aq) + 8H^+(aq)],[Mn^{2+}(aq) + 4H_2O(l)]\,|\,Pt$$

and use the standard electrode potentials to confirm that the reaction is energetically feasible:

Redox process	$E^{\ominus}$/V
$MnO_4^-(aq) + 8H^+(aq) + 5e^- \rightleftharpoons Mn^{2+}(aq) + 4H_2O(l)$	+1.51
$Fe^{3+}(aq) + e^- \rightleftharpoons Fe^{2+}(aq)$	+0.77
$MnO_4^-(aq) + 8H^+(aq) + 5Fe^{2+}(aq)$	
$\rightarrow Mn^{2+}(aq) + 4H_2O(l) + 5Fe^{3+}(aq)$	+0.74

$E^{\ominus}$ is positive (+0.74 V) so the reaction is energetically feasible.

Homogeneous catalysis

Many aqueous transition metal ions act as homogeneous catalysts. We might expect the reaction between peroxodisulphate(VI) ions and iodide ions to be slow because both are anions. The reaction is catalysed by iron(II) cations or iron(III) cations. You could explain how the catalyst works by taking part in the redox reaction to provide an alternative route with a lower activation energy.

Iron(III) ions rapidly oxidize the iodide ions to iodine:

$$2Fe^{3+}(aq) + 2I^-(aq) \rightarrow 2Fe^{2+}(aq) + I_2(aq) \qquad E^{\ominus} = 0.23\ V$$
$$2Fe^{2+}(aq) + S_2O_8^{2-}(aq) \rightarrow 2Fe^{3+}(aq) + 2SO_4^{2-}(aq) \qquad E^{\ominus} = 1.24\ V$$

Iron(II) ions are rapidly oxidized by the peroxodisulphate(IV) ions.
Combine the two steps involving the catalyst to get the equation

complex ion formation

$$S_2O_8^{2-}(aq) + 2I^-(aq) \rightarrow 2SO_4^{2-}(aq) + I_2(aq)$$

The catalyst takes part in the reaction but is not used up.

Complex formation

→ A complex ion is a central metal cation with six (four or two) ligands (molecules or ions) datively bonded to it.

→ The nucleophilic molecules CO, H_2O, NH_3 and nucleophilic ions OH^-, CN^-, F^-, Cl^- are monodentate ligands.

→ 1,2-diaminoethane $NH_2CH_2CH_2NH_2$ and the ethanedioate ion $^-O_2C \cdot CO_2^-$ are bidentate ligands.

→ EDTA, ethylenediaminetetraacetic acid, is a hexadentate ligand consisting of many atoms, six of which are each capable of forming one dative bond with the central ion.

The shapes of complex ions may sometimes (not always) be predicted using VSEPR theory. Octahedral is most common and can give rise to geometrical and optical isomeric complexes. There are also linear, square planar and tetrahedral complexes.

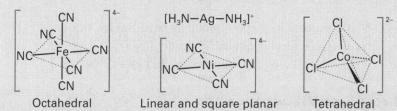

Octahedral   Linear and square planar   Tetrahedral

→ Reactions can occur with one ligand displacing another

When you dissolve copper(II) carbonate in the minimum amount of concentrated hydrochloric acid you get a dirty yellow-brown solution containing $[CuCl_4]^{2-}$ anions. When you add water, the solution changes colour from yellow-brown through green to blue, the colour of an aqueous solution containing $[Cu(H_2O)_6]^{2+}$ cations.

→ Water molecules can replace chloride ions as ligands:

$$[CuCl_4]^{2-}(conc \cdot HCl) + 6H_2O(l) \rightleftharpoons [Cu(H_2O)_6]^{2+}(aq) + 4Cl^-(aq)$$

If you add ammonia to aqueous copper(II) sulphate, the alkali first precipitates copper(II) hydroxide which redissolves in excess ammonia to form a deep blue solution.

→ Ammonia molecules can replace water molecules as ligands:

$$[Cu(H_2O)_6]^{2+}(aq) + 4NH_3(aq) \rightleftharpoons [Cu(NH_3)_4(H_2O)_2]^{2+}(aq) + 4H_2O(l)$$

Exam question (12 min)   answer: page 134

(a) Explain, with an example in each case, the meaning of the term
   (i) complex cation, (ii) complex anion, (iii) bidentate ligand.

(b) The antidote for cyanide poisoning is to freshly mix $Na_2CO_3(aq)$ with
   $FeSO_4(aq)$ containing citric acid (2-hydroxypropane-1,2,3-trioic acid) and
   swallow immediately.
   (i) Explain why the two solutions cannot be mixed and stored until needed
      and suggest how the antidote works.
   (ii) Draw the structural formula of citric acid and suggest *two* reasons why
      it is in the antidote.

The jargon

An *anion* is a negatively charged ion.
A *cation* is a positively charged ion.

The jargon

A *ligand* is a nucleophile (molecule or anion) datively bonded by a lone electron-pair to a central transition metal cation.
Monodentate means one tooth or dative bond:

$$EDTA = \begin{array}{l} CH_2N(CH_2CO_2H)_2 \\ | \\ CH_2N(CH_2CO_2H)_2 \end{array}$$

Links

See pages 15 and 17: VSEPR theory.

Examiner's secret

You get more marks if your drawing shows clearly the three-dimensional shape of a complex.

1st transition series: chromium

The jargon

Greek *chroma* – colour

Action point

Find out the colours of the following
insoluble chromates:
silver chromate, Ag_2CrO_4
lead(II) chromate, $PbCrO_4$
barium chromate, $BaCrO_4$

Watch out!

Cr has four oxidation numbers if you
include 0 for the oxidation state of the
uncombined metal.

The jargon

Chromium trioxide is the old name for CrO_3.

Watch out!

If you add $AgNO_3$(aq) to $K_2Cr_2O_7$(aq) you
will get a precipitate of silver chromate
(*not* dichromate).

Chromium is a typical d-block element and transition metal. Many chromium compounds are brightly coloured. Some are used in paints because the colours do not fade in the sun.

The element

Chromium shows all the general characteristics of a transition metal:

→ high melting point (second highest in the first transition series)
→ hard, lustrous and resistant to chemical attack (chromium plating)
→ forms tough alloy with iron (stainless steel)

The Goldschmidt process used aluminium to reduce chromium oxide to the metal in a type of 'thermite' reaction:

$$Cr_2O_3 + 2Al \rightarrow 2Cr + Al_2O_3$$

Oxidation states

Chromium exists in three oxidation states:

→ +2   strongly reducing (spontaneously oxidized by air)
→ +3   most stable as chromium(III) oxide, Cr_2O_3
→ +6   strongly oxidizing (as dichromate(VI) ion, $Cr_2O_7^{2-}$)

Chromium(VI) compounds

Chromium(VI) oxide

Chromium(VI) oxide forms when concentrated sulphuric acid is added to cold, concentrated aqueous sodium dichromate(VI). Red crystals of CrO_3(s) separate from the orange sodium dichromate solution.

$$Cr_2O_7^{2-}(aq) + H_2SO_4(l) \rightarrow 2CrO_3(s) + SO_4^{2-}(aq) + H_2O(l)$$

→ Chromium(VI) oxide is a powerful oxidizing agent

Before modern surfactants, 'chromic acid' (a solution of chromium(VI) oxide in concentrated sulphuric acid) was used for cleaning laboratory glassware. It is extremely corrosive, it reacts violently with many organic compounds and it causes skin ulcers. These days chemists prefer to clean glassware with safer modern surfactants.

Chromates and dichromates

Na_2CrO_4 and $K_2Cr_2O_7$ strictly refer to sodium chromate(VI) and potassium dichromate(VI) but we often leave off the (VI).

→ The oxidation state of chromium is +6 in chromates *and* dichromates.
→ Sodium dichromate is more soluble than potassium dichromate.
→ Aqueous chromate solutions are *yellow* and stable when *alkaline or neutral*.
→ Aqueous dichromate solutions are *orange* and stable when *acidic*.
→ Chromate and dichromate ions are related by the acid–base reaction
$2CrO_4^{2-}(aq) + 2H^+(aq) \rightleftharpoons Cr_2O_7^{2-}(aq) + H_2O(l)$.

This equilibrium explains why insoluble chromates precipitate when some aqueous metal ions are added to aqueous dichromate ions.

Chromium(III) compounds ●●●

Chromium(III) oxide and hydroxide

→ amphoteric oxide prepared *in a fume cupboard* by starting (heat) the exothermic 'decomposition' of ammonium dichromate:

$$(NH_4)_2Cr_2O_7 \rightarrow Cr_2O_3 + 4H_2O + N_2$$

→ amphoteric hydroxide precipitated by aqueous NaOH(aq) from aqueous chromium(III) salts as a green jelly that dissolves in excess NaOH(aq) to form a green solution.

Chromium(II) compounds ●●●

Obtained by reducing chromium(III) compounds (e.g. Cr^{3+}(aq) → Cr^{2+}(aq) by acid and zinc amalgam) and keeping the chromium(II) product in an inert atmosphere or under reducing conditions.

Isomerism in chromium compounds ●●●

Hydration isomerism

Hydrated chromium(III) chloride, $CrCl_3 \cdot 6H_2O$, can exist as isomers depending upon whether the chlorine is ionic or is a ligand.

$[Cr(H_2O)_4Cl_2]^+Cl^- \cdot 2H_2O$	dark green
$[Cr(H_2O)_6]^3(Cl^-)_3$	grey-blue
$[Cr(H_2O)_5Cl]^{2+}(Cl^-)_2 \cdot H_2O$	light green

Geometrical isomerism

→ octahedral dichlorotetraam-mine chromium(III) ions can exist as geometrical (*cis* and *trans*) isomers

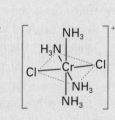

Optical isomerism

→ 1,2-diaminoethane ligands can complex chromium(III) to give optical (*mirror-image*) isomers

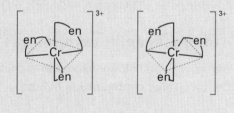

Watch out!

Decompositions are endothermic. The dichromate ion oxidizes the ammonium ion to nitrogen exothermically! This reaction is very dangerous because dichromate dust becomes airborne!

The jargon

Latin: *cis* – on this side of
trans – on the other side of

Checkpoint

Draw diagrams to show geometric isomers of the square planar complex, $Pt(NH_3)_2Cl_2$.

The jargon

en = $H_2NCH_2CH_2NH_2$ (bidentate ligand)

Exam question (8 min)   answer: page 134

(a) Deduce the oxidation state of chromium in (i) $Cr_2(SO_4)_3$ and (ii) CrO_2Cl_2.

(b) Construct equations for (i) the reaction of aqueous sodium hydroxide with $Cr_2(SO_4)_3$(aq), (ii) the formation of chromium dichloride dioxide from NaCl(s), $K_2Cr_2O_7$(s) and H_2SO_4(l), (iii) the hydrolysis of chromium dichloride dioxide by NaOH(aq) to form aqueous chromate(VI) ions.

1st transition series: manganese

For A-level you focus on manganese compounds in which the element shows a wide range of oxidation states. The +2, +4 and +7 states are most important but you may also meet the +6 state.

Manganese(VII) compounds

→ For A-level the most important compound of manganese is potassium manganate(VII), $KMnO_4$.

→ Potassium manganate(VII) is used as an oxidizing agent in titrimetric (volumetric) analysis.

We cannot prepare directly (by dissolving $KMnO_4(s)$ crystals in water) solutions of precise concentration and they deteriorate on standing especially in the presence of dust and other organic material. So

→ $MnO_4^-(aq)$ is not used as a primary standard.

We prepare chlorine in the laboratory fume cupboard by putting drops of concentrated hydrochloric acid onto $KMnO_4(s)$ crystals.

$$2KMnO_4 + 16HCl \rightarrow 2KCl + 2MnCl_2 + 8H_2O + 5Cl_2$$

Manganese(IV) compounds

→ Manganese(IV) oxide, MnO_2, occurs naturally as the mineral *pyrolusite* and is the only common manganese(IV) compound.

The black oxide is formed when manganese(II) nitrate decomposes on heating or when $MnO_4^{2-}(aq)$ disproportionates.

→ Hydrated manganese(IV) oxide (brown) may form if insufficient acid is present when aqueous manganate(VII) is reduced:
$$MnO_4^-(aq) + 4H^+(aq) + 3e^- \rightarrow MnO_2(s) + 2H_2O(l)$$
$$+7 \qquad\qquad\qquad -3 \quad\; +4$$

→ Manganese(IV) oxide oxidizes concentrated hydrochloric acid to chlorine: $MnO_2(s) + 4HCl(aq) \rightarrow Cl_2(g) + MnCl_2(aq) + 2H_2O(l)$.

→ $MnO_2(s)$ catalyses the decomposition of aqueous hydrogen peroxide: $2H_2O_2(aq) \rightarrow 2H_2O(l) + O_2(g)$.

Manganese(II) compounds

→ +2 is the most stable oxidation state of manganese.

→ Electronic configuration of Mn^{2+} is $1s^2 2s^2 2p^6 3s^2 3p^6 3d^5$.

Manganese(II) oxide

→ basic oxide

→ made by heating manganese(II) ethanedioate (compare FeO):
$$MnC_2O_4(s) \rightarrow MnO(s) + CO(g) + CO_2(g)$$

The reducing atmosphere of carbon monoxide prevents oxidation. The colour of the manganese(II) oxide is greyish green.

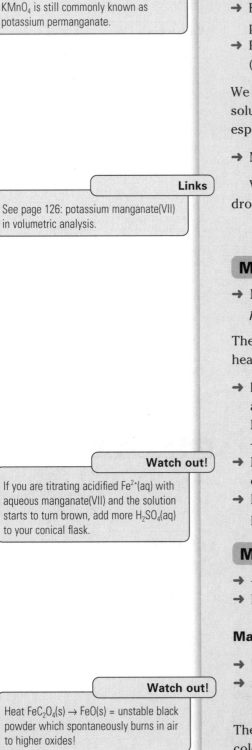

The jargon

$KMnO_4$ is still commonly known as potassium permanganate.

Links

See page 126: potassium manganate(VII) in volumetric analysis.

Watch out!

If you are titrating acidified $Fe^{2+}(aq)$ with aqueous manganate(VII) and the solution starts to turn brown, add more $H_2SO_4(aq)$ to your conical flask.

Watch out!

Heat $FeC_2O_4(s) \rightarrow FeO(s)$ = unstable black powder which spontaneously burns in air to higher oxides!

Manganese(II) hydroxide

→ Formed as a white precipitate when NaOH(aq) is added to aqueous manganese(II) sulphate:

$$Mn^{2+}(aq) + 2OH^-(aq) \rightarrow Mn(OH)_2(s)$$

The precipitate does not dissolve in excess alkali.

→ White $Mn(OH)_2(s)$ precipitate gradually turns brown as it oxidizes to manganese(III) oxide.

Manganese(II) sulphate

→ Hydrated manganese(II) sulphate crystals have a faint pink colour.

This salt forms a variety of crystalline hydrates. $MnSO_4 \cdot H_2O$, $MnSO_4 \cdot 4H_2O$ and $MnSO_4 \cdot 7H_2O$ are all known.

Reactions of the aqueous manganese(II) ion

You could investigate the reactions of aqueous manganese(II) ions using aqueous manganese(II) sulphate.

	Reagent	Result
1	$NaOH(aq)$ or $NH_3(aq)$	white precipitate formed at first, $Mn(OH)_2(s)$, gradually darkens to manganese(III) oxide
2	$Na_3PO_4(aq)$ and $NH_3(aq)$	pink precipitate of manganese(II) ammonium phosphate
3	$PbO_2(s)$ and conc. $HNO_3(aq)$	pink/purple colour of manganate(VII) ions produced by oxidation of Mn^{2+}
4	$NaBiO_3(s)$ stirred into cold $HNO_3(aq)$ containing $Mn^{2+}(aq)$	pink/purple colour of manganate(VII) ions observed on standing or filtering

Equations for tests 3 and 4:

$$5PbO_2(s) + 2Mn^{2+}(aq) + 4H^+(aq)$$
$$\rightarrow 2MnO_4^-(aq) + 5Pb^{2+}(aq) + 2H_2O(l)$$
$$5NaBiO_3(s) + 2Mn^{2+}(aq) + 14H^+(aq)$$
$$\rightarrow 2MnO_4^-(aq) + 5Bi^{3+}(aq) + 5Na^+(aq) + 7H_2O(l)$$

→ Lead(IV) oxide in nitric acid and sodium bismuthate(V) in nitric acid can oxidize manganese(II) to manganate(VII).

Exam question (7 min) answer: page 134

(a) How many moles of each of the following would be needed to reduce one mole of manganate(VII) ions to manganese(II) ions in excess acid?

   (i) iron(II) ions

   (ii) ethanedioate ions

   (ili) hydrogen peroxide molecules

(b) (i) Write a balanced ionic equation for the oxidation of $SO_3^{2-}(aq)$ to $SO_4^{2-}(aq)$ by manganate(VII) ions which are reduced to manganese(II) ions in excess acid. (ii) 20.0 cm³ of a solution containing $SO_3^{2-}(aq)$ ions react with 16.0 cm³ of aqueous potassium manganate(VI) of concentration 0.015 mol dm⁻³, in acidic solution. Calculate the concentration of $SO_3^{2-}(aq)$ ions in mol dm⁻³.

Checkpoint 1

Write the formula for manganese(III) oxide and suggest an equation for its formation from $Mn(OH)_2$.

Action point

Make a list of hydroxides precipitated by NaOH(aq) from aqueous transition metal ions and note whether or not they dissolve in excess alkali.

Checkpoint 2

Outline how you would prepare a sample of manganese(II) carbonate from manganese(II) sulphate.

The jargon

Manganese(II) ammonium phosphate = $NH_4MnPO_4 \cdot H_2O(s)$.

Watch out!

Sodium bismuthate(V) is an extremely sensitive test for $Mn^{2+}(aq)$. You need to stir only a little of the solid into the acidified (with HNO_3) solution you are testing.

1st transition series: copper

Examiner's secret

You should know the names of SO_2, NO, NO_2 and N_2O_4. In equation 3 N_2O_4 would be accepted instead of $2NO_2$.

The jargon

Latin *verdigris* = green of Greece.

The jargon

$CuSO_4 \cdot 5H_2O$ is also called copper(II) sulphate-5-water.

We can tell copper from all other metals by its colour. The blue colour of hydrated copper(II) sulphate crystals and solution is another distinguishing feature of this element.

The element   ●●●

The standard electrode potential for Cu^{2+}/Cu is +0.34 volt. For the other nine metals in the first transition series from Sc to Zn the value of $E^{\ominus}$ is negative, so copper is the least reactive of these ten elements.

→ Copper does not react with acids like HCl(aq) to give hydrogen.
→ Copper reacts with oxidizing acids to form copper(II) compounds and various acid reduction products depending on the conditions

1 $Cu(s) + 2H_2SO_4(\text{hot conc.}) \rightarrow CuSO_4(aq) + 2H_2O(l) + SO_2(g)$
2 $3Cu(s) + 8HNO_3(\text{dil.}) \rightarrow 3Cu(NO_3)_2(aq) + 4H_2O(l) + 2NO(g)$
3 $Cu(s) + 4HNO_3(\text{conc.}) \rightarrow Cu(NO_3)_2(aq) + 2H_2O(l) + 2NO_2(g)$

When copper metal is exposed to the atmosphere, it gradually becomes covered with a greenish-blue layer (*verdigris*). The layer is usually a basic form of copper(II) carbonate, chloride or sulphate but its composition varies from place to place, e.g. by the sea it is mainly copper(II) chloride. When heated in air or oxygen copper becomes coated with black copper(II) oxide: $2Cu(s) + O_2(g) \rightarrow 2CuO(s)$.

Copper(II) compounds   ●●●

Copper(II) oxide, CuO

→ Basic oxide prepared in the laboratory by heating copper(II) carbonate or nitrate: $2Cu(NO_3)_2 \rightarrow 2CuO + 2NO_2 + O_2$

Copper(II) hydroxide, $Cu(OH)_2$

→ Precipitated from aqueous copper(II) salts by aqueous ammonia or NaOH(aq): $Cu^{2+}(aq) + 2OH^-(aq) \rightarrow Cu(OH)_2(s)$

Its slight solubility in *excess concentrated* sodium hydroxide indicates that this basic oxide has some amphoteric character.

Copper(II) carbonate

→ Precipitated from aqueous copper(II) sulphate by $Na_2CO_3(aq)$ (highly alkaline) as a *basic carbonate*, e.g. $2CuCO_3 \cdot Cu(OH)_2(s)$

Basic carbonates vary in composition and occur naturally in ores, e.g. *azurite* $2CuCO_3 \cdot Cu(OH)_2$ and *malachite* $CuCO_3 \cdot Cu(OH)_2$.

Aqueous copper(II) cation

Copper(II) oxide, hydroxide or carbonate react with *excess* $H_2SO_4(aq)$ to give a blue (acidic) solution from which we can obtain blue crystals of copper(II) sulphate pentahydrate, $CuSO_4 \cdot 5H_2O$. Four water ligands are attached to each copper ion; the other molecule of water is *hydrogen bonded* within the lattice.

→ The complex ion $[Cu(H_2O)_6]^{2+}$ has a distorted octahedral structure with two water molecules at its apices and the other four ligands (square) coplanar with the central cation.

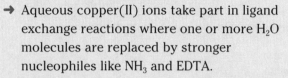

→ Aqueous copper(II) ions take part in ligand exchange reactions where one or more H_2O molecules are replaced by stronger nucleophiles like NH_3 and EDTA.

When you add drops of aqueous ammonia to copper(II) salt solutions, you can expect to see a pale blue precipitate (copper(II) hydroxide) form and then dissolve in excess ammonia to give a deep blue solution containing tetraamminecopper (II) ions.

$$[Cu(H_2O)_6]^{2+}(aq) + 4NH_3(aq) \rightleftharpoons [Cu(H_2O)_2(NH_3)_4]^{2+}(aq) + 4H_2O(l)$$

When you add drops of concentrated hydrochloric acid to aqueous copper(II) sulphate, you can expect the solution to change from blue to green.

$$\underset{\text{blue}}{[Cu(H_2O)_6]^{2+}(aq)} + 4Cl^-(aq) \rightleftharpoons \underset{\text{yellow/brown}}{[CuCl_4]^{2-}(aq)} + 6H_2O(l)$$

If you dilute the green solution with water, the blue colour returns.

→ The tetrachlorocuprate(II) ion, $[CuCl_4]^{2-}(aq)$, has a tetrahedral shape.

Copper(I) compounds ●●○

Copper(I) oxide

→ Cu_2O forms as a red solid when aldehydes and sugars reduce Fehling's solution (a *complex* Cu^{2+} ion in alkaline conditions).

The solid reacts with acids to form copper metal and copper(II) salts:
$Cu_2O(s) + 2H^+(aq) \rightarrow Cu(s) + Cu^{2+}(aq) + H_2O(l)$ (disproportionation)

Copper(I) halides

→ CuCl is prepared as a white solid by reducing copper(II) chloride in hot concentrated HCl with copper and pouring the mixture into water: $Cu(s) + CuCl_2(conc \cdot HCl) \rightarrow 2CuCl(conc \cdot HCl) \rightarrow 2CuCl(s)$
→ CuBr is prepared in a similar way but copper(I) iodide is formed when aqueous copper(II) ions oxidize aqueous iodide ions:
$2Cu^{2+}(aq) + 4I^-(aq) \rightarrow 2CuI(s) + I_2(aq)$

The iodine may be measured by titrating with aqueous sodium thiosulphate using starch indicator.

→ Aqueous copper(I) ions spontaneously disproportionate unless stabilized as a complex, e.g. $[CuCl_2]^-(aq)$ and $[CuCl_4]^{3-}(aq)$.

> **Watch out!**
>
> In Fehling's solution Cu^{2+} is complexed with a bidentate ligand to stop $Cu(OH)_2$ precipitating in the alkaline conditions, and not all sugars are reducing agents!

Exam question (8 min)   answer: page 134

(a) Describe and explain what happens to aqueous copper(II) chloride on adding (i) iron filings, (ii) aqueous ammonia until in excess, (iii) sulphur dioxide.

(b) Calculate the volume of $0.100 \, mol \, dm^{-3}$ HCl(aq) which will react with $0.05 \, mol$ of basic copper(II) carbonate: $2CuCO_3 \cdot Cu(OH)_2$.

Structured exam question

answer: pages 135

This question concerns patterns in the properties of the hydrides, oxides and chlorides of elements shown on the right in a portion of the periodic table.

Li	Be	B	C
Na	Mg	Al	Si	P	S	Cl	Ar
			Ge
			Sn
			Pb

(a) Using only those elements listed above,

   (i) give the formula of an ionic hydride:

   ....

   ....

   (ii) give the formula of a diatomic covalent hydride and explain why it is polar:

   ....

   ....

   (iii) give the formula of a non-polar covalent hydride and explain why it is non-polar:

   ....

   ....

   (iv) give the equation for the reaction of a diatomic hydride with water to form an alkaline solution:

   ....

   ....

(b) State how the acid–base character of the oxides changes across the period from sodium to chlorine and illustrate your statement by writing an equation for an appropriate reaction for each of the following oxides: statement:

   ....

   ....

   sodium oxide:

   ....

   sulphur dioxide:

   ....

(c) Describe a *trend* in the properties, *down the group* from carbon to lead, of

   (i) the oxides CO_2, SiO_2, SnO_2, PbO_2:

   ....

   ....

   (ii) the chlorides CCl_4, $SiCl_4$, $SnCl_4$, $PbCl_4$:

   ....

   ....

(d) Compare the reactions of sodium chloride, aluminium chloride and silicon chloride with water to illustrate how the ease of hydrolysis varies across the period Na to Ar.

   ....

   ....

(e) Write an equation for the action of heat upon (i) lithium carbonate, (ii) magnesium carbonate and suggest why the other group I carbonates do not readily decompose.

   ....

   ....

(20 min)

Answers
Inorganic chemistry

Patterns and trends in the periodic table

Checkpoints

1 Noble (inert) gases, group 0
2 (a) $MgO + 2H^+ \rightarrow Mg^{2+} + H_2O$
   (b) $CO_2 + OH^- \rightarrow HCO_3^-$
   (c) (i) $Al_2O_3 + 6H^+ \rightarrow 2Al^{3+} + 3H_2O$
      (ii) $Al_2O_3 + 2OH^- + 3H_2O \rightarrow 2[Al(OH)_4]^-$

Exam question

(a) (i) Aluminium oxide
   (ii) Germanium
   (iii) Lead(IV) oxide
(b) Oxides become less acidic with increasing atomic number. The +2 oxidation state becomes more stable
All valid answers score marks when there is more than one possible answer.

Groups I and II: alkali and alkaline earth metals

Checkpoints

1 (a) $Mg + H_2O \xrightarrow{steam} MgO + H_2$
   (b) $Ba + 2H_2O \rightarrow Ba(OH)_2 + H_2$
2 (a) (i) Hard water is water that does not readily lather with soap since it contains magnesium and/or calcium ions which react with soaps to form an insoluble scum.
      (ii) Limescale is calcium carbonate formed by the decomposition of hydrogencarbonate ions in temporarily hard water.
      $$Ca^{2+} + 2HCO_3^- \rightarrow CaCO_3 + CO_2 + H_2O$$
   (b) Stalactites and stalagmites form when water evaporates – solid $Ca(HCO_3)_2$ is too unstable to exist:
      $$Ca^{2+} + 2HCO_3^- \rightarrow CaCO_3 + CO_2 + H_2O$$
   (c) Small highly charged ions distort the symmetry of the carbonate ion so that it decomposes to an oxide ion and carbon dioxide. Although the calcium ion is doubly charged the smaller lithium ion has the greater polarizing power and lithium carbonate decomposes at a lower temperature than calcium carbonate.

Exam question

(a) (i) $CaO + H_2O \rightarrow Ca(OH)_2$
   (ii) $BaO_2 + 2H_2O \rightarrow Ba(OH)_2 + H_2O_2 \rightarrow H_2O + \frac{1}{2}O_2$ [in hot water]
   (iii) $LiH + H_2O \rightarrow LiOH + H_2$
(b) (i) $3Mg + N_2 \rightarrow Mg_3N_2$
   (ii) $TiCl_4 + 2Mg \rightarrow 2MgCl_2 + Ti$
   (iii) $NO_3^- + 4Mg + 10H^+ \rightarrow 4Mg^{2+} + 3H_2O + NH_4^+$

Industrial chemistry: s-block

Checkpoints

1 (a) (i) Anode (ii) Cathode
   (b) This is to separate the products.
   (c) Membrane cells are more efficient and environmentally friendly. In the mercury cathode cell, small amounts of toxic mercury escape into the environment and in the diaphragm cells some chlorine is lost.
2 (a) $2NaCl + CaCO_3 \rightarrow CaCl_2 + Na_2CO_3$
   (b) (i) Ammonia is very volatile and very soluble; therefore there will always be some losses.
      (ii) Changes in industrial practice are always made on the basis of economics. The use of a carbonate mineral may also present fewer environmental problems which is reflected in costs.
3 (a) Chlorine corrodes steel.
   (b) To separate the products of electrolysis.
   (c) Some calcium may be discharged at the cathode which will alloy with the sodium.
4 (a) $CaCO_3 \rightarrow CaO + CO_2$
   (b) (i) A reducing agent
      (ii) Calcium   +2 to zero
         Aluminium   zero to +3

Exam question

(a) (i) When magnesium is heated in air it burns with a bright white flame, forming both the oxide and nitride:
      $$Mg + \tfrac{1}{2}O_2 \rightarrow MgO$$
      $$3Mg + N_2 \rightarrow Mg_3N_2$$
   (ii) A reaction occurs but only the nitride reacts:
      $$Mg_3N_2 + 6H_2O \rightarrow 3Mg(OH)_2 + 2NH_3$$
   (iii) The red litmus turns blue.
(b) (i) Magnesium carbonate and lithium carbonate both decompose on heating.
   (ii) Beryllium oxide and aluminium oxide are both amphoteric.

Group III: aluminium and boron

Checkpoints

1 Boron $1s^2 2s^2 2p^1$
   Aluminium $1s^2 2s^2 2p^6 3s^2 3p^1$
2 (a) Ammonia will be pyramidal (three bonding pairs and one non-bonding pair of electrons).
   (b) $H_3N{-}BCl_3$ will be tetrahedral around both the nitrogen atom and the boron atom as each is surrounded by four bonding pairs of electrons.
   (c) The tetrahydridoborate(III) ion is tetrahedral as there are four bonding pairs of electrons in the boron valence shell.
3 The white precipitate is aluminium hydroxide and the colourless gas is carbon dioxide. Sodium carbonate (the salt of a strong base and weak acid) is hydrolysed and so has an alkaline solution. The aqueous aluminium ion is acidic.
   $$2Al^{3+} + 3Na_2CO_3 + 3H_2O \rightarrow 2Al(OH)_3 + 3CO_2 + 6Na^+$$

Exam question

When aqueous ammonia is added to aqueous aluminium sulphate, a white precipitate is formed which is insoluble in excess of the reagent:
$$Al^{3+}(aq) + 3OH^-(aq) \rightarrow Al(OH)_3(s)$$
When aqueous ammonia is added to aqueous zinc sulphate, a white precipitate is formed which is soluble in excess aqueous ammonia to form a colourless solution:
$$Zn^{2+}(aq) + 2OH^-(aq) \rightarrow Zn(OH)_2(s)$$
$$Zn(OH)_2(s) + 4NH_3(aq) \rightarrow [Zn(NH_3)_4]^{2+} + 2OH^-(aq)$$

Group IV: elements and oxides

Checkpoints

1 Metals are ductile, malleable and sonorous.
2 A, Diamond; B, graphite; C, fullerene, C_{60}.
3 (a) Iron is more reactive than tin so the iron corrodes (rusts) when the metals are exposed to moist air:
$$Fe \rightarrow Fe^{2+} + 2e^-$$
Zinc is more reactive than iron so the zinc is oxidized when the metals are exposed to moist air:
$$Zn \rightarrow Zn^{2+} + 2e^-$$
These electrons suppress the oxidation of iron.
  (b) Lead was used in the production of leaded ('antiknock') petrol. Phasing out this fuel caused a decrease in demand for lead.
4 (a) Silicon dioxide and silicates are refractory materials (unaffected by very high temperatures).
  (b) The atoms in silicon dioxide are held firmly in the giant (tetrahedral) lattice structure by strong covalent bonds (compare diamond). Thus the compound acts as an abrasive by removing atoms from other structures without losing atoms from its own structure.

Exam question

(a) When carbon dioxide is passed into limewater, a white precipitate of calcium carbonate is observed:
$$Ca(OH)_2 + CO_2 \rightarrow CaCO_3 + H_2O$$
Because calcium hydroxide is only sparingly soluble, the carbon dioxide quickly removes all the calcium and hydroxide ions. Excess carbon dioxide then reacts with carbonate ions to form aqueous hydrogencarbonate ions and the white precipitate dissolves to give a colourless solution:
$$CO_2 + CO_3^{2-} + H_2O \rightarrow 2HCO_3^-$$
(b) In the carbonate ion, the central carbon is bonded to three equivalent oxygen atoms. VSEPR theory predicts a trigonal planar ion with more than one arrangement of bonds. So the structure has delocalized electrons (dotted lines).

Group IV: chlorides and hydrides

Checkpoints

1 (a) Lead(II) nitrate(V); (b) plumbate(II) ion.
2 (a) You might think it is energetically unfavourable (see pages 27–40) but $\Delta H^{\ominus} = -139$ kJ mol for liquid

tetrachloromethane. So it must be kinetically unfavourable with an activation energy so high that the reaction does not take place at an appreciable rate. (b) Aqueous iodine is only a very dilute solution of I_2 (0.0013 mol dm^{-3} at 25 °C) and just perceptibly yellow/brown. The covalent iodine molecule is very much more soluble in covalent organic solvents. When shaken with tetrachloromethane (which is immiscible with water and more dense), almost all the iodine dissolves in the CCl_4 giving a violet colour and leaving the aqueous layer colourless.

3 Assuming the HCl is gaseous, then

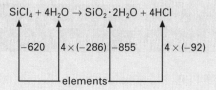

$$SiCl_4 + 4H_2O \rightarrow SiO_2 \cdot 2H_2O + 4HCl$$
-620 4 × (−286) −855 4 × (−92)

elements

Hence $\Delta H = -855 - 368 + 620 + 1144 = +541$ kJ mol^{-1}
4 Hexachlorostannate(IV) ion; hexachloroplumbate(IV) ion

Exam question

(a) (i) $Si(s) + 2Cl_2 \rightarrow SiCl_4(l)$ Heat silicon powder in dry chlorine gas and collect product in a dry receiver.
   (ii) $Pb(s) \rightarrow Pb^{2+}(aq) + 2Cl^-(aq) \rightarrow PbCl_2(s)$ Dissolve lead powder in nitric acid in a fume cupboard, add sodium chloride to the aqueous lead(II) nitrate and filter off the white precipitate.
(b) (i) The chloride is violently hydrolyzed to SiO_2 and NaCl.
   (ii) The chloride dissolves to form a solution containing the tetrachloroplumbate(II) anion $[PbCl_4]^{2+}$.

Group V: elements and oxides

Checkpoints

1 Oxygen and the noble gases, mainly argon but all the others as well.
2 Covalent radius decreases L → R across a period so a boron atom is bigger than a nitrogen atom. Hence ♠ represents boron.
3 Glowing splint thermally decomposes dinitrogen oxide into a mixture of 33% oxygen 67% nitrogen. This concentration of oxygen is enough to re-light the splint.

Exam question

(a) (i) $2NO + O_2 \rightarrow 2NO_2$. Nitrogen dioxide is a brown gas
   (ii) $2N_2O \rightarrow 2N_2 + O_2$ Glowing splint thermally decomposes dinitrogen oxide into a mixture of 33% oxygen 67% nitrogen. This concentration of oxygen is enough to re-light the splint.
(b) (iii) Sodium nitrite $NaNO_2$,
   (ii) Sodium nitrate $NaNO_3$,
   (iii) Sodium phosphate(III) Na_3PO_3,
   (iv) Sodium phosphate(V) Na_3PO_4.

Group V: oxoacids and hydrides

Checkpoints

1 $N_2O_4 + 2NaOH \rightarrow NaNO_3 + NaNO_2 + H_2O$

2 (a) $pH = -\log_{10}(0.02) = 1.7$

(b) Use a titrimetric method to find the volume of nitric acid which will exactly neutralize a given volume of aqueous sodium hydroxide. Repeat the experiment using exactly the same volumes of acid and alkali but omit the indicator. Evaporate the resulting solution to small bulk and allow the salt to crystallize.

3 $H_2PO_4^-$

4 $PCl_5 + H_2O \rightarrow POCl_3 + 2HCl$

5 (a) When heated ammonium chloride undergoes thermal dissociation:
$$NH_4Cl(s) \rightleftharpoons NH_3(g) + HCl(g)$$
On the cooler sides of the tube the products recombine to form the white solid ammonium chloride.

(b) However, the ammonia molecule is lighter than the hydrogen chloride molecule and so diffuses faster. Ammonia gas reaches the end of the test tube before hydrogen chloride gas and the alkaline gas turns moist red litmus paper blue:
$$NH_3(g) + H_2O(l) \rightleftharpoons NH_4^+(aq) + OH^-(aq)$$

Exam question

(a) $N_2O_4 + 2NaOH \rightarrow NaNO_2 + NaNO_3 + H_2O$ Oxidn nos. +3 and +5

(b) (i) 0.1 mol PCl_3 reacts with 0.3 mol H_2O. Molar mass H_2O is 18.0 g mol^{-1}. So mass water needed is 0.3 mol × 18.0 g mol^{-1} = 5.4 g.

(ii) 0.1 mol PCl_3 could give 0.3 mol HCl or 0.3 × 24 dm^3 = 7.2 dm^3 at room temperature and pressure.

Group VI: oxygen and sulphur

Checkpoint

$KNO_3 \rightarrow KNO_2 + {}^1\!/_2O_2$

$HgO \rightarrow Hg + {}^1\!/_2O_2$

$2Pb_3O_4 \rightarrow 6PbO + O_2$

You would get the marks for any other correct examples.

Exam question

(a) (i) When sulphur is heated it becomes a mobile amber liquid as the S_8 rings break away from their positions in the solid sulphur lattice and become free to move in the liquid state. As the temperature rises S_8 rings break open and the liquid darkens because of unpaired electrons. The viscosity of the liquid increases as broken rings combine into longer sulphur atom chains which interact with one another. The decrease in the mobility of the liquid reaches a maximum (around 180 °C). As the temperature rises to the boiling point of 444 °C, the long chains begin to break up and the liquid becomes more mobile (smaller molecules) and much darker (more unpaired electrons).

(ii) When the sulphur is poured into cold water, the chains of sulphur atoms persist, giving the rubber-like

form known as plastic sulphur which only slowly reverts to the S_8 structure of rhombic sulphur.

(b) (i) $Mg + S \rightarrow MgS$ (ii) $Cl_2 + 2S \rightarrow S_2Cl_2$

(c) Sulphur has an oxidation state of −2 in the sulphide ion since it is far more electronegative than sodium. In covalent sulphur hexafluoride, fluorine is the more electronegative and forces sulphur into an oxidation state of +6.

Group VI: water and hydrogen peroxide

Checkpoints

1 (a) H_2O

(b) O^{2-}

In $H_2O(l) + H_2O(l) \rightleftharpoons H_3O^+(aq) + OH^-(aq)$ water acts as both a Brønsted–Lowry acid (proton donor) and a Brønsted–Lowry base (proton acceptor).

2 An ion-exchange resin softens hard water by removing calcium ions and replacing them with sodium ions:
$$[resin(2Na^+)_n] + Ca^{2+}(aq) \rightarrow [resin(Ca^{2+})_n] + 2Na^+(aq)$$
Eventually the resin runs out of sodium ions. Salt provides sodium ions to regenerate the resin by displacing the calcium ions it removed from the water:
$$[resin(Ca^{2+})_n] + 2Na^+(aq) \rightarrow [resin(2Na^+)_n] + Ca^{2+}(aq)$$

Exam question

The two ion–electron half-equations are
$$MnO_4^- + 8H^+ + 5e^- \rightarrow Mn^{2+} + 4H_2O$$
$$2H_2O_2 \rightarrow 2H_2O + O_2 + 2e^-$$
Therefore 2 mol $MnO_4^- \equiv 5$ mol H_2O_2

No. of mol of MnO_4^- = (37.5 × 10^{-3} × 0.018 7)
= 7.01 × 10^{-4} mol

No. of mol H_2O_2 in 25.0 cm^3 of the solution
= 2.5 × 7.01 × 10^{-4} mol

No. of mol H_2O_2 in 1.00 dm^3
$$= 2.5 \times 7.01 \times 10^{-4} \text{ mol} \times \frac{1\,000}{25.0}$$
$$= 0.070\,1 \text{ mol}$$

Concentration of the aqueous hydrogen peroxide = 0.070 1 mol dm^{-3}

Group VII: halogens and hydrogen halides

Checkpoints

1 (a) In the reaction with water $Cl_2 + H_2O \rightarrow HClO + HCl$ the oxidation state of chlorine changes from zero to +1 *and* −1. Disproportionation is the simultaneous oxidation and reduction of the same element.

(b) All covalent non-metal chlorides formed by direct combination of the element with chlorine are hydrolysed by water and many metal chlorides, e.g. aluminium chloride and iron(III) chloride, formed by direct combination are also hydrolysed. So we must use anhydrous reagents and keep apparatus dry.

131

2 (a) $NH_3 + HCl \rightarrow NH_4Cl$

(b) $2KMnO_4 + 16HCl \rightarrow 2MnCl_2 + 2KCl + 5Cl_2 + 8H_2O$

Exam question

(a) (i) $2NaOH + Cl_2 \rightarrow NaCl + NaClO + H_2O$ −1 and +1

(ii) $6NaOH + 3Cl_2 \rightarrow 5NaCl + NaClO_3 + 3H_2O$ −1 and +5

(b) The HF_2^- is linear (two bonding pairs of electrons and no non-bonding pairs).

Group VII: halides and interhalogen compounds

Checkpoints

1 When sulphuric acid reacts with HBr the change in oxidation state of sulphur is +6 to +4; with HI the change is +6 to −2. Therefore, HI is the stronger reducing agent.

2 $NaHCO_3 + HCl \rightarrow NaCl + H_2O + CO_2$

3 (a) (i) In alkaline solution, silver oxide, Ag_2O, is precipitated.

(ii) Nitric acid must be used since hydrochloric would produce a precipitate of silver chloride, and sulphuric acid, a precipitate of silver sulphate.

(b) The silver halides become increasingly covalent in character.

(c) The lattice enthalpy of calcium fluoride is large enough for it to be insoluble.

4 (a) Iodine chloride

(b) Iodine trichloride

Exam question

(a) (i) $2Al + 3Cl_2 \rightarrow Al_2Cl_6$

(ii) $2MnO_4^- + 16H^+ + 10I^- \rightarrow 5I_2 + 2Mn^{2+} + 8H_2O$

(b) One mole of iodate(V) ions liberates three moles of iodine. Therefore $0.15 \text{ mol } I_2$ is liberated by $0.05 \text{ mol } IO_3^-$.

Concentration $= \left(\dfrac{1\,000}{20}\right) \times 0.05 \text{ mol dm}^{-3} = 2.5 \text{ mol dm}^{-3}$

Industrial chemistry: aluminium and carbon

Checkpoint

Compound readily hydrolysed because the central aluminium atom (unlike boron in $NaBH_4$) has d-orbitals available to accept a lone pair of electrons from a water molecule. (Compare with water on CCl_4 and $SiCl_4$.)

Exam question

(a) (i) Filtration. (ii) Inefficient electrolysis and impure product. (iii) Some cryolite is lost because the reactions in the cell are complex and involve fluoride ions.

(b) Aqueous sulphuric acid is electrolysed and oxygen liberated at the aluminium anode reacts with the metal surface to form aluminium oxide. Anodizing thickens the oxide layer on aluminium surfaces to adsorb dyes and permanently colour the metal surface.

Industrial chemistry: silicon and nitrogen

Checkpoints

1 The oxidation state is +4.

2 Asbestos dust can enter the body and damage the lungs. Blue asbestos is very dangerous and listed as a carcinogen causing asbestosis, a disease which develops to produce a variety of carcinoma.

3 Unconverted hydrogen and nitrogen are recycled.

Exam question

(a) Process converts atmospheric nitrogen into a form in which it can be assimilated by plants.
Those conditions of temperature and pressure most favourable to producing the most economic yield of ammonia.
A catalyst in a *different* phase from the reactants.

(b) $$\frac{p^2_{NH_3}}{p_{N_2} \times p^3_{H_2}} = K_p$$

(i) If at a given temperature (K_p is constant) the pressure is increased, then the denominator would momentarily increase. To bring K_p back to its constant value the partial pressure of NH_3 must increase. Therefore an increase in pressure favours the formation of ammonia.

(ii) Cooling will favour formation of ammonia – reaction is exothermic so K_p increases as temperature falls. Removing ammonia encourages more nitrogen and hydrogen to combine trying to reach equilibrium.

(c) (i) Convenient, economic and likely to produce a finely divided form of iron with a large surface area.

(ii) $Fe_2O_3 + 3H_2 \rightarrow 2Fe + 3H_2O$

Industrial chemistry: sulphur and the halogens

Checkpoints

1 (a) (i) Impurities like arsenic oxide poison the catalyst and make it less efficient.

(ii) The percentage conversion would decrease.

(b) (i) A catalyst which is in a different phase from the reactants.

(ii) A promoter is a substance which when added to a catalyst improves the catalytic effect.

2 $$\frac{p^2_{SO_3}}{p^2_{SO_2} \times p_{O_2}} = K_p$$

The expression for K_p from the law of chemical equilibrium shows that its units are $(\text{pressure})^{-1}$.
If at a given temperature (K_p is constant) the pressure is increased, then the denominator would momentarily increase. To bring K_p back to its constant value the partial pressure of SO_3 must increase. Therefore an increase in pressure favours the formation of sulphur trioxide.

As the forward reaction is exothermic, an increase in temperature decreases the value of K_p and therefore the yield of sulphur trioxide decreases.

3 CFCs are chlorofluorocarbons – very stable compounds which have been used as aerosol propellants and refrigerants. They can persist in the atmosphere, reach the stratosphere and deplete the ozone layer by taking part in free radical reactions. The EU has banned their manufacture in order to allow the ozone layer to recover.

Exam question

(a) Sulphur dioxide and excess air are purified and passed over a heterogeneous catalyst of promoted vanadium(V) oxide at an optimum temperature of about 430 °C:
$$2SO_2(g) + O_2(g) \rightleftharpoons 2SO_3(g)$$
Sulphur trioxide gas is absorbed in 98% sulphuric acid to form oleum ($H_2S_2O_7$). The oleum is diluted to form 100% sulphuric acid (H_2SO_4).

(b) (i) Colourless steamy fumes evolve. Volatile hydrogen chloride displaced by less volatile sulphuric acid:
$$H_2SO_4 + NaCl \rightarrow NaHSO_4 + HCl.$$
(ii) Brown fumes condensing to drops of pale yellow fuming liquid. Volatile nitric acid displaced by less volatile sulphuric acid: $H_2SO_4 + KNO_3 \rightarrow KHSO_4 + HNO_3$. Colour caused by thermal decomposition of nitric acid into brown nitrogen dioxide:
$$2HNO_3 \rightarrow 2NO_2 + H_2O + \tfrac{1}{2}O_2.$$
(iii) Purple vapour condensing to shiny black crystals, some steamy fumes and smell of bad eggs. Sulphuric acid displaces hydrogen iodide which thermally decomposes and also reduces sulphuric acid to hydrogen sulphide:
$$H_2SO_4 + KI \rightarrow KHSO_4 + HI \cdot 2HI \rightarrow H_2 + I_2$$
$$8HI + H_2SO_4 \rightarrow H_2S + 4H_2O + 4I_2$$
(iv) Fruity pear drop smell produced. Sulphuric acid catalyzes formation of pentyl ethanoate ester:
$$CH_3CO_2H + C_5H_{11}OH \rightleftharpoons CH_3CO_2C_5H_{11} + H_2O$$

(c) (i) Sea water; (ii) sodium iodate in Chile saltpetre.

(d) Sea water is acidified and the bromide ions oxidized to bromine by chlorine in a non-metal displacement reaction: $Cl_2 + 2Br^- \rightarrow 2Cl^- + Br_2$.
Sodium iodate(V) is reduced to iodine using sulphur dioxide: $5SO_2 + 2KIO_3 + 4H_2O \rightarrow I_2 + 4H_2SO_4 + K_2SO_4$

(e) In the presence of dilute nitric acid, aqueous silver nitrate produces a white precipitate with aqueous chloride ions which is soluble in aqueous ammonia, a cream precipitate with aqueous bromide ions which is partially soluble in aqueous ammonia, a primrose yellow precipitate with aqueous iodide ions which is insoluble in aqueous ammonia, but no precipitate with aqueous fluoride ions since silver fluoride is soluble.

(f) Hydrochloric acid for cleaning iron and steel surfaces prior to galvanizing, spray-painting and tin-plating. Chloroethene used as the monomer in the production of poly(chloroethene) or PVC.

Group VIII (or 0): the noble gases

Checkpoints

1 Noble metals are very unreactive d-block metals below hydrogen in the electrochemical series.
Rh   Pd   Ag
Ir   Pt   Au

2 The energy required to bring about
$$O_2(g) \rightarrow O_2^+(g) + e^-$$

Exam question

(a) Ar $1s^2 2s^2 2p^6 3s^2 3p^6$
A chlorine atom Cl $1s^2 2s^2 2p^6 3s^2 3p^5$ gains an electron to become a chloride anion Cl⁻ $1s^2 2s^2 2p^6 3s^2 3p^6$ and a potassium atom K $1s^2 2s^2 2p^6 3s^2 3p^6 4s^1$ loses an electron to become a potassium cation K⁺ $1s^2 2s^2 2p^6 3s^2 3p^6$. The chloride ion, potassium ion and argon atom are isoelectronic.

(b) Pass a dry mixture of xenon and fluorine gases over hot metallic nickel, acting as a catalyst, in the absence of air and moisture.
XeF_2: Here we have three non-bonding pairs of electrons and two bonding pairs. This gives a linear structure.
XeF_4: Here there are four bonding pairs of electrons and two non-bonding pairs giving a square planar structure:

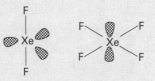

represents a non-bonding pair of electrons

1st transition series: metals, aqueous ions and redox

Checkpoints

1 (a) $TiCl_4 + 4Na \rightarrow Ti + 4NaCl$
(b) $Ni + 4CO \rightleftharpoons Ni(CO)_4$
(c) $Cu^{2+} + 2e^- \rightarrow Cu$

2 (a)

$ScCl_3$	$TiCl_4$	VCl_5	$CrCl_3$	$MnCl_2$
$FeCl_3$	$CoCl_2$	$NiCl_2$	$CuCl_2$	$ZnCl_2$

(b)

$Sc_2(SO_4)_3$	$Fe_2(SO_4)_3$
$Ti(SO_4)_2$	$CoSO_4$
$V_2(SO_4)_5$	$NiSO_4$
$Cr_2(SO_4)_3$	$CuSO_4$
$MnSO_4$	$ZnSO_4$

Exam question

(a) The electronic configuration is Zn $1s^2 2s^2 2p^6 3s^2 3p^6 3d^{10} 4s^2$ and the d-shell electrons show zinc to be a d-block element. The d-shell is always full, so the oxidation state of zinc in its compounds is +2 and, unlike the transition metals, does not vary.

(b) (i) Orange to green to blue as the oxidation state of chromium is reduced from +6 to +3 to +2; (ii) yellow to blue to green to violet as the oxidation state of vanadium is reduced from +5 to +4 to +3 to +2.

133

1st transition series: redox reactions and complex ion formation

Checkpoint

Questions on free energy do not apply to every examination board.

$\Delta G^{\oplus} = -5 \times 96\,500 \times 0.74\ \mathrm{J\ mol}^{-1} = -357\,000\ \mathrm{J\ mol}^{-1}$

Exam question

(a) (i) A complex cation is a positively charged ion consisting of a central metal ion with molecules or ions (ligands) datively bonded to it. $[Cu(NH_3)_4]^{2+}$

   (ii) A complex anion is a negatively charged ion consisting of a central metal ion with molecules or ions (ligands) datively bonded to it. $[CuCl_4]^{2-}$

   (iii) A bidentate ligand is a molecule or anion capable of forming two dative bonds with a central metal ion. 1,2-diaminoethane $NH_2CH_2CH_2NH_2$

(b) (i) The carbonate would neutralize the citric acid, make the solution alkaline and the iron would precipitate as a basic hydroxide. The iron(II) cations would remove poisonous cyanide ions by forming a stable non-poisonous complex anion, $[Fe(CN)_6]^{4-}$

   (ii)
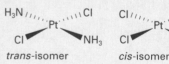

The acid lowers the pH of the solution and suppresses the hydrolysis of the iron(II) sulphate. It reacts with the sodium carbonate to form carbon dioxide which makes the mixture fizzy and easier to swallow.

1st transition series: chromium

Checkpoint

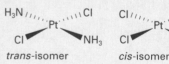

trans-isomer   *cis*-isomer

Exam question

(a) (i) +3 because charge on the sulphate ion is 2− so charge on the chromium ion must be 3+;

   (ii) +6 because oxidation state of oxygen is −2 and chlorine is −1.

(b) (i) $6NaOH(aq) + Cr_2(SO_4)_3(aq) \rightarrow 3Na_2SO_4(aq) + 2Cr(OH)_3(s)$

   (ii) $4NaCl + K_2Cr_2O_7 + 3H_2SO_4 \rightarrow 2CrO_2Cl_2 + 2Na_2SO_4 + K_2SO_4 + 3H_2O$

   (iii) $4NaOH + CrO_2Cl_2 \rightarrow Na_2CrO_4(aq) + 2NaCl(aq) + 2H_2O(l)$

1st transition series: manganese

Checkpoints

1 $3Mn(OH)_2 \rightarrow Mn_2O_3 + 3H_2O$

2 Add a solution of sodium carbonate to aqueous manganese(II) sulphate and filter off the precipitated carbonate. Wash, dry and store.

$MnSO_4(aq) + Na_2CO_3(aq) \rightarrow MnCO_3(s) + Na_2SO_4(aq)$

Try writing this ordinary equation as an ionic equation.

Exam question

(a) (i) 5 (ii) 2.5 (iii) 2.5

(b) (i) $2MnO_4^- + 6H^+ + 5SO_3^{2-} \rightarrow 2Mn^{2+} + 3H_2O + 5SO_4^{2-}$

   (ii) $0.015 \times \left(\dfrac{5}{2}\right) \times \left(\dfrac{16.0}{20.0}\right) = 0.030\ \mathrm{mol\ dm}^{-3}$

1st transition series: copper

Exam question

(a) (i) The blue colour of the solution fades and a reddish precipitate of copper forms:

$Cu^{2+}(aq) + Fe(s) \rightarrow Fe^{2+}(aq) + Cu(s)$

   (ii) A blue precipitate forms which dissolves in excess aqueous ammonia to form a deep blue solution:

$Cu^{2+}(aq) + 2OH^-(aq) \rightarrow Cu(OH)_2(s)$

With excess ammonia

$Cu(OH)_2(s) + 4NH_3(aq) \rightarrow [Cu(NH_3)_4]^{2+}(aq) + 2OH^-(aq)$

   (iii) With sulphur dioxide, a reducing agent, insoluble copper(I) chloride is formed as a white precipitate:

$2CuCl_2 + SO_2 + 2H_2O \rightarrow 2CuCl + H_2SO_4 + 2HCl$

(b) One mole of basic copper(II) carbonate produces two moles of carbonate ion and two moles of hydroxide ion.

$2CO_3^{2-} + 4H^+ \rightarrow 2CO_2 + 2H_2O$

$2OH^- + 2H^+ \rightarrow 2H_2O$

One mole of basic copper(II) carbonate requires six moles of $H^+ \equiv$ six moles of HCl.

0.05 mol basic copper(II) carbonate $\equiv$ 0.30 mol of $H^+ \equiv 3.0\ \mathrm{dm}^3$ of HCl of concentration $0.100\ \mathrm{mol\ dm}^{-3}$.

Examiner's secrets

You might gain more marks if you gave the following equations:

$[Cu(H_2O)_6]^{2+}(aq) + 2OH^-(aq) \rightleftharpoons [Cu(H_2O)_4(OH)_2](s) + 2H_2O(l)$

$[Cu(H_2O)_6]^{2+}(aq) + 4NH_3(aq) \rightleftharpoons [Cu(H_2O)_2(NH_3)_4]^{2+}(aq) + 4H_2O(l)$

and you referred to the equilibria moving to the right as ammonia is added and the tetraamminediaquocopper(II) ion formed.

Incidentally, all six water ligands are replaced if the reaction is in liquid ammonia instead of water.

Structured exam question

(a) (i) Na^+H^-

(ii) $H—Cl$

because chlorine is more electronegative than hydrogen so the two atoms share the one pair of electrons unequally.

(iii) CH_4

because the electronegativities of carbon and hydrogen are similar and the tetrahedral shape of the molecule gives a net dipole moment of zero.

(iv) $Na^+H^-(s) + H_2O(l) \rightarrow Na^+(aq) + OH^-(aq) + H_2(g)$

Examiner's secrets

Remember that group I metals are powerful reducers and give up electrons to form ionic compounds.

(b) Statement: oxides of Na and Mg are basic, aluminium oxide is amphoteric and the oxides of Si, P, S and Cl are acidic.

sodium oxide: $Na_2O(s) + H_2O(l) \rightarrow 2Na^+(aq) + 2OH^-(aq)$

sulphur dioxide: $SO_2(g) + H_2O(l) \rightarrow H_2SO_3(aq)$

(c) (i) The oxides change from acidic (CO_2, SiO_2) to amphoteric (SnO_2, PbO_2) and become oxidizing agents (SnO_2, PbO_2).

(ii) From CCl_4 to $PbCl_4$ the chlorides show a decrease in thermal stability and an increase in the tendency to disproportionate into the divalent state, giving off chlorine gas.

(d) Sodium chloride simply dissolves in water. Aluminium chloride fumes in moist air dissolves exothermically in water to give an acidic solution:

$$AlCl_3(s) + 6H_2O(l) \rightarrow [Al(H_2O)_5OH]^+(aq) + H^+(aq)$$

Examiner's secrets

You may get a bonus mark for an answer that is a bit special.

Silicon tetrachloride reacts violently with water to give dense fumes of hydrogen chloride and a gelatinous precipitate of hydrated silicon oxide (silica gel):

$$SiCl_4(l) + 4H_2O \rightarrow SiO_2 \cdot 2H_2O + 4HCl$$

(e) (i) $Li_2CO_3 \rightarrow Li_2O + CO_2$ (ii) $MgCO_3 \rightarrow MgO + CO_2$

The polarizing power of the very small lithium cation is high enough to distort and promote the decomposition of the carbonate anion. The polarizing power of the other larger group I cations is not high enough to do this.

Get used to structural formulae and you will make sense of organic chemistry. You already know H_2O (H—O—H), CO_2 (O=C=O) and therefore that hydrogen, oxygen and carbon form respectively one, two and four covalent bonds. So you could deduce the formula and structure of ethyne, the simplest hydrocarbon, as C_2H_2 (H—C≡C—H). Could you work out the structure of ethene (C_2H_4) and ethane (C_2H_6)? You must also get used to nomenclature like *eth*ane, *eth*ene, *eth*yne. How many carbon atoms in *eth*anol? You will learn about modern instruments that tell us the structure of compounds. You will also learn how we simplify our study of organic chemistry by classifying compounds, reagents (as free radicals, electrophiles and nucleophiles) and types of reaction (as addition, elimination and substitution).

Organic chemistry

Exam themes

→ Give the name, structural formula and typical properties of hydrocarbons and of members of homologous series of oxygen, nitrogen and halogen containing compounds
→ Describe types of isomerism, including *cis–trans* and optical isomerism, and be able to work out possible isomers for a given molecular formula
→ Describe and explain the typical reactions of homologous series in terms of the structure and properties of functional groups
→ Describe reactions as free radical, electrophilic addition, nucleophilic substitution and be able to write equations and state conditions for reactions
→ Describe modern spectroscopic techniques and use IR, UV, NMR and mass spectra to identify and determine the structure of a particular compound
→ Describe some of the principles and processes involved in the industrial production of organic chemicals and polymers

Topic checklist

O AS ● A2

	AQA	CCEA	EDEXCEL	OCR	WJEC
How to name compounds	O	O	O	O	O
Classes of compounds and functional groups	O	O	O	O	O
Types of reactions and reagents	O	O	O	O	O
Hydrocarbons: alkanes and alkenes	O	O	O	O	O
Hydrocarbons: alkenes and arenes	O	O	O	O	O
Compounds containing halogens	O●	O●	O●	O●	O●
Compounds containing oxygen: alcohols, phenols, aldehydes and ketones	O●	O●	O●	O●	O●
Compounds containing oxygen: aldehydes, ketones and carboxylic acids	O●	O●	O●	O●	O●
Compounds containing nitrogen: amines and amino acids	O●	O●	O●	O●	O●
Compounds containing nitrogen: proteins and polyamides	O●	O●	O	O●	O●
Instrumental techniques: UV, visible and IR spectroscopy	●	●	●	●	●
Instrumental techniques: NMR and mass spectrometry	●	●	O●	●	●
Industrial chemistry: oil refining and petrochemicals	O●	O●		O●	O●

How to name compounds

Chemists try to give organic compounds unambiguous names that tell us about the structure of the substance. We use a set of nomenclature rules to generate systematic names that may consist of prefix(es), root and suffix together with numbers and punctuation.

Naming alkanes

Many organic compounds have a root derived from the names of alkanes.

Study these five rules.

→ Base the root name on the longest continuous carbon chain and the straight-chain alkane with the same number of carbons.
→ Add prefixes based on the shorter carbon branches and the names of the corresponding straight-chain alkanes.
→ Indicate the number of identical branches by adding di- (two), tri- (three), tetra- (four), etc.
→ Number the positions of the branches on the longest chain from the end giving the lower number for the initial branching point.
→ Attach the prefixes in alphabetical order of branch name.

Examples

$$CH_3CHCH_2CH_3$$ with CH_3 branch

2-methylbutane

$$CH_3CH_2CHCH—CHCH_2CH_3$$ with CH_3 CH_3 branches

3,4-dimethylheptane

Naming alkenes

We name alkenes in the same way as alkanes but with the following important differences:

→ Base the root name on the straight-chain alkane with the same number of C-atoms as the longest continuous carbon chain that contains the double bond.
→ Use a number to show the position of the double bond.
→ Use *cis*- and *trans*- as prefixes to name simple geometric isomers with the same group on each of two doubly bonded carbon atoms.

Examples

$$H_2C=CHCH_2CH_3$$

but-1-ene

$$CH_3C=C—CH_3$$ with CH_3 branch

2-methylbut-2-ene

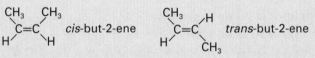

cis-but-2-ene

trans-but-2-ene

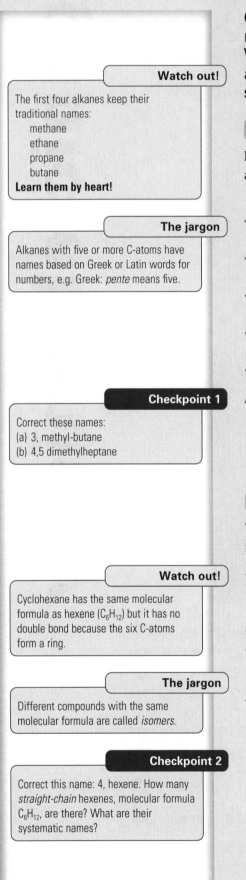

Watch out!

The first four alkanes keep their traditional names:
   methane
   ethane
   propane
   butane
Learn them by heart!

The jargon

Alkanes with five or more C-atoms have names based on Greek or Latin words for numbers, e.g. Greek: *pente* means five.

Checkpoint 1

Correct these names:
(a) 3, methyl-butane
(b) 4,5 dimethylheptane

Watch out!

Cyclohexane has the same molecular formula as hexene (C_6H_{12}) but it has no double bond because the six C-atoms form a ring.

The jargon

Different compounds with the same molecular formula are called *isomers*.

Checkpoint 2

Correct this name: 4, hexene. How many *straight-chain* hexenes, molecular formula C_6H_{12}, are there? What are their systematic names?

Naming arenes

Arenes are hydrocarbons containing benzene rings.

→ The simplest arene (C_6H_6) is called benzene.
→ The group (C_6H_5—) derived from it is called phenyl.
→ The carbon atoms are numbered to show the position of any groups attached to the ring.

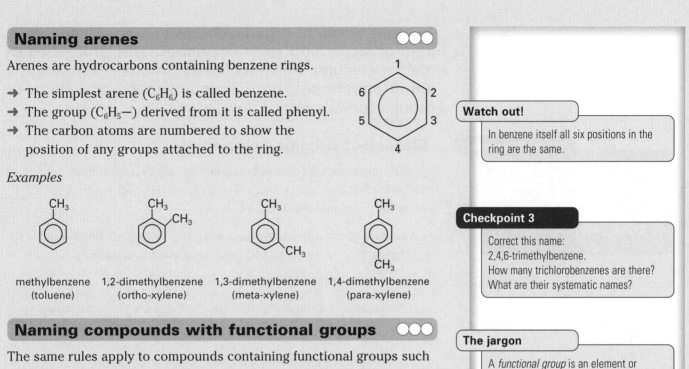

Examples

methylbenzene (toluene)   1,2-dimethylbenzene (ortho-xylene)   1,3-dimethylbenzene (meta-xylene)   1,4-dimethylbenzene (para-xylene)

Naming compounds with functional groups

The same rules apply to compounds containing functional groups such as halogeno (—Cl, —Br, —I), hydroxyl (—OH) and amino (—NH_2).

→ Name the functional groups in alphabetical order.

Example

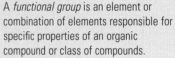

2,4-dibromo-2-chloro-3-methylhexane

Watch out!

In benzene itself all six positions in the ring are the same.

Checkpoint 3

Correct this name: 2,4,6-trimethylbenzene. How many trichlorobenzenes are there? What are their systematic names?

The jargon

A *functional group* is an element or combination of elements responsible for specific properties of an organic compound or class of compounds.

Action point

Make yourself a set of cards to learn the names, structures and reactions of the important functional groups listed on pages 140–1.

Exam question   answer: page 167

(a) Explain what is meant by the statement that 2-methylbutane and 2,2-dimethylpropane are *isomeric alkanes* and draw the structures of the two hydrocarbons.

(b) Give the systematic name for *each* of the following:

(i)
$$CH_3CHCH_3CHCH_3$$
with CH_3 and CH_3 substituents

(ii) $CH_3CH=CHCH_2CH_3$

(iii) a benzene ring with CH_3, two Cl groups and Br

(c) Draw structures for each of the following:

(i) octane, (ii) cyclopentane, (iii) 3-chloroethylbenzene.

(d) Draw the three possible structures for the molecular formula $C_2H_2Cl_2$ and give the systematic name of each structure.

Classes of compounds and functional groups

To make sense of organic chemistry, we classify compounds according to the structure and properties of their molecules and functional groups. Here are the important classes, homologous series and functional groups for A-level.

Classes of organic compounds

We put compounds into broad classes (e.g. *aliphatic* and *aromatic*) which we divide into narrower classes (e.g. *hydrocarbons, oxygen-containing, nitrogen-containing*, etc.).

→ A homologous series is the simplest class of organic compound.
→ Alkanes are the simplest and most important homologues.

Alkanes (general formula C_nH_{2n+2})

CH_4	C_2H_6	C_3H_8	C_4H_{10}	C_5H_{12} ...
methane	ethane	propane	butane	pentane

→ Homologues have the same functional group(s).

Aliphatic primary alcohols (general formula $C_nH_{2n+1}OH$)

CH_3OH	C_2H_5OH	C_3H_7OH	C_4H_9OH	$C_5H_{11}OH$ ...
methanol	ethanol	propan-1-ol	butan-1-ol	pentan-1-ol

Functional groups in hydrocarbons

Functional group		Prefix	Suffix	Class of compounds
>C:C<	>C=C<		ene	alkenes
C_6H_5-	⬡−	phenyl	benzene	arenes and aromatics

Functional groups containing oxygen

Functional group		Prefix	Suffix	Class of compounds
−CHO	−C(H)(=O)		al	aldehydes
>CO	>C=O	oxo	one	ketones and aldehydes
$-CO_2H$	−C(=O)(O−H)		oic acid	carboxylic acids
−COCl	−C(=O)(Cl)		oyl chloride	acid or acyl chlorides
$-CO_2$	−C(=O)(O−)		oate	esters and polyesters
$(-CO)_2O$	−C(=O)(O)(=O)C−		anhydride	acid anhydrides

Functional group	Prefix	Suffix	Class of compounds
$-CH_2OH$	$-\overset{H}{\underset{H}{C}}-O{-}H$	ol	primary alcohols
$>CHOH$	$-\overset{}{\underset{H}{C}}-O{-}H$	ol	secondary alcobols
$>COH$	$-\overset{}{\underset{}{C}}-O{-}H$	ol	tertiary alcohols

Functional groups containing nitrogen ●●●

Functional group	Prefix	Suffix	Class of compounds
$-NH_2$	$-N\begin{smallmatrix}H\\H\end{smallmatrix}$	amino	amine	primary amines α-amino acids
$-CONH_2$	$-C\begin{smallmatrix}O\\N\!H\\H\end{smallmatrix}$	amido	amide	amides
$-NO_2$	$-N\begin{smallmatrix}O\\O\end{smallmatrix}$	nitro		nitro compounds
$-CN$	$-C\equiv N$	cyano	nitrile	nitriles

→ Proteins contain $-CO-NH-$ and are built from α-amino acids.

The jargon

α-amino acid = $\underset{NH_2}{-\overset{\gamma}{C}-\overset{\beta}{C}-\overset{\alpha}{C}-CO_2H}$

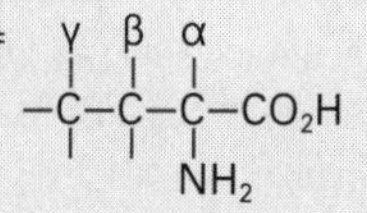

α-, β-, γ- label the C-atoms attached to the carboxyl functional group

The jargon

$-CO-NH-$ is an amide linkage

Functional groups containing halogen ●●●

Functional group	Prefix	Suffix	Class of compounds
-hal	-hal	halogeno		halogeno compounds
$-COCl$	$-C\begin{smallmatrix}O\\Cl\end{smallmatrix}$		oyl chloride	acid or acyl chlorides

The jargon

hal = F, Cl, Br, I

Exam question (8 min)   answer: page 167

(a) Write down *all* the functional groups you can identify in the compound known as CS, the structure of which is shown below.

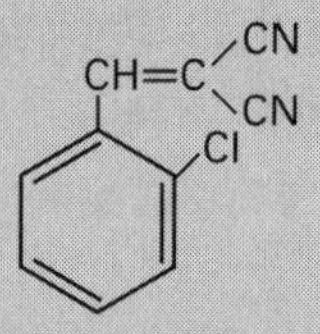

$$CH{=}C\begin{smallmatrix}CN\\CN\\Cl\end{smallmatrix}$$

(b) Draw the structures of
   (i) ethanoic acid, (ii) ethanol, (iii) ethyl ethanoate.

(c) Construct a balanced equation for the reaction of ethanol and ethanoic acid.

Examiner's secrets

Here is a typical question you've been warned about. You may have heard about CS gas being used by the police but you aren't expected to have come across its formula. You can answer this question by applying your knowledge and understanding of structural formulae and functional groups.

Use the tables on these two pages to answer part (a).

141

Types of reactions and reagents

There are four types of organic reaction: addition, elimination, rearrangement and substitution. They may involve polar bonds and electrons moving around. To understand the mechanism of a reaction, you must know about electrophiles, nucleophiles and free radicals; so what are they?

Types of reagent ●●●

Free radicals

When a single covalent bond (R—R) between two atoms with similar electronegativities breaks *homolytically*, each atom keeps one electron. The resulting reactive molecular fragments (R·) are called free radicals, e.g. ·Cl and ·CH_3.

→ A free radical is an atom (or group of atoms) with an unpaired electron.

Electrophiles and nucleophiles

When a single covalent bond (A—B) between two atoms with different electronegativities breaks *heterolytically*, one atom (A) loses both electrons and the other atom (:B) keeps both electrons. Fragment A is an electrophile, e.g. nitronium ion NO_2^+. Fragment :B is a nucleophile, e.g. hydroxide ion :OH^-.

→ An electrophile is an electron-pair acceptor (and a Lewis acid).
→ A nucleophile is an electron-pair donor (and a Lewis base).

Types of reaction ●●●

Free radical substitution

The reaction between an alkane and bromine in UV or strong sunlight is a chain reaction:

→ initiation $Br—Br \rightarrow 2 \cdot Br$
→ propagation $\cdot Br + H—CH_3 \rightarrow H—Br + \cdot CH_3$
          $Br—Br + \cdot CH_3 \rightarrow Br—CH_3 + \cdot Br$

The net effect of these two propagation steps is given by

$$Br—Br + H—CH_3 \rightarrow H—Br + Br—CH_3$$

→ termination $\cdot CH_3 + \cdot CH_3 \rightarrow CH_3—CH_3$

CH_3Br can take part in the propagation steps to form CH_2Br_2. Substitution can continue until all four H-atoms in the methane molecule have been substituted.

Electrophilic addition

Propene reacts with hydrogen chloride to form 2-chloropropane:

Secondary carbocation

Checkpoint 1

Identify the following as free radical, electrophile or nucleophile.
(i) $AlCl_3$ (ii) Br (iii) NH_3 (iv) C_2H_4

Checkpoint 2

Write the names and structural formulae of the organic compounds formed when methane is brominated.

The jargon

CFCs are *chlorofluorocarbons* formerly used as aerosol propellants and refrigerants. They form chlorine radicals in the stratosphere.
Ozone, O_3, is readily attacked by radicals, e.g. $\cdot Cl + O_3 \rightarrow O_2 + \cdot ClO$

The jargon

A curly ⌢ arrow shows the movement of a pair of electrons and must be used correctly. The tail shows where the electron-pair comes from. The head shows where the electron-pair is going to.

The high electron density of the C=C bond makes alkenes react as nucleophiles and induces a polarity in approaching halogen molecules.

→ Addition occurs across the double bond to the two carbon atoms originally joined by it.
→ The addition product is 2-chloropropane, *not* 1-chloropropane, because the reaction follows Markovnikov's rule: when a molecule HZ adds to an unsymmetrical alkene, the H-atom adds to the double-bonded C-atom with the greater number of H-atoms.

Nucleophilic substitution

1-bromobutane is hydrolysed slowly in *one step* by aqueous sodium hydroxide to form butan-1-ol.

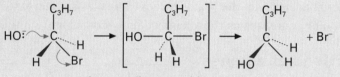

Intermediate transition state

As a lone pair of electrons on the hydroxide ion attacks the region of low electron density and begins to form a bond with the electrophilic carbon atom attached to the halogen, the carbon–halogen bond begins to break. In the middle of this process an unstable transition state is reached.

→ Primary halogenoalkanes favour the S_N2 mechanism.

2-bromo-2-methylpropane is hydrolysed rapidly in *two steps* by aqueous sodium hydroxide to form 2-methylpropan-2-ol.

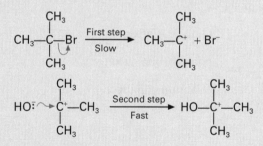

→ Tertiary halogenoalkanes favour the S_N1 mechanism

Exam question (10 min)   answer: page 167

(a) For the photochemical chlorination of ethane, write balanced equations to show (i) the initiation reaction, (ii) *two* propagation reactions, (iii) a termination reaction, and (iv) the *overall* reaction showing $C_2H_4Cl_2$ as the organic product.

(b) Suggest why the frequency of light needed for free radical halogenation of alkanes is higher for chlorine than for bromine.

(c) Suggest how a CFC such as CCl_2F_2 might destroy ozone and explain why depletion of the ozone layer is considered harmful.

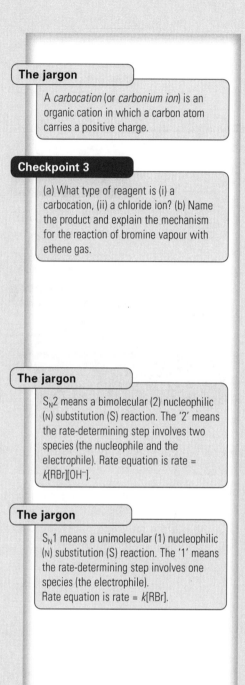

The jargon

A *carbocation* (or *carbonium ion*) is an organic cation in which a carbon atom carries a positive charge.

Checkpoint 3

(a) What type of reagent is (i) a carbocation, (ii) a chloride ion? (b) Name the product and explain the mechanism for the reaction of bromine vapour with ethene gas.

The jargon

S_N2 means a bimolecular (2) nucleophilic (N) substitution (S) reaction. The '2' means the rate-determining step involves two species (the nucleophile and the electrophile). Rate equation is rate = $k[RBr][OH^-]$.

The jargon

S_N1 means a unimolecular (1) nucleophilic (N) substitution (S) reaction. The '1' means the rate-determining step involves one species (the electrophile).
Rate equation is rate = $k[RBr]$.

Watch out!

The CH_3 group in methylbenzene reacts like methane with free radicals.

143

Hydrocarbons: alkanes and alkenes

Hydrocarbons are compounds of carbon and hydrogen only. They may be saturated or unsaturated and their molecules may contain chains and/or rings of carbon atoms.

The jargon

Saturated means there are only single C—C bonds. *Unsaturated* means there are multiple bonds (e.g. double C=C) in chains and rings.

Checkpoint

Predict and draw the shape of an ethane molecule.

Examiner's secrets

To see if you understand basic ideas you may be asked to suggest why the b.p. of $C(CH_3)_4$ (= 9.5 °C) is lower than b.p. of $CH_2CH_2CH_2CH_2CH_3$ (= 36 °C). Revise van der Waals' forces on page 18.

Alkanes

→ Simplest homologous series ($C_nH_{(2n+2)}$)

The first four are petroleum gases and methane is also natural gas. Alkanes are also constituents of diesel, petrol, aviation fuel, oils, etc.

→ Nomenclature of the other homologous series is based on alkanes

The melting and boiling points of the alkanes increase as molar mass increases (because van der Waals forces increase as n increases). Boiling point decreases as the alkane chain becomes more branched.

Reactions of alkanes

The alkanes undergo very few reactions.

Combustion

Alkanes can burn to water and carbon dioxide (or carbon monoxide if insufficient air): $CH_4 + 2O_2 \rightarrow CO_2 + 2H_2O$.

Chlorination

→ Alkanes undergo substitution reactions with chlorine in the presence of UV light.
→ UV light causes homolytic fission of chlorine molecules and the formation of free radicals.

1 $Cl_2 + h\upsilon \rightarrow 2Cl\cdot$ chain initiation
2 $Cl\cdot + CH_4 \rightarrow \cdot CH_3 + HCl$
3 $\cdot CH_3 + Cl_2 \rightarrow CH_3Cl + Cl\cdot$ } chain propagation
4 $\cdot CH_3 + \cdot CH_3 \rightarrow C_2H_6$ chain termination
5 Further substitution can give CH_2Cl_2, $CHCl_3$ and CCl_4.

The jargon

Thermal cracking is by heat. *Catalytic cracking* is by heat and heterogeneous catalysts.

Cracking

Smaller alkane molecules are more useful than some of the larger ones obtained from the primary distillation of petroleum.

→ Alkanes are split by thermal and catalytic cracking into smaller alkane molecules and some unsaturated molecules:

$$C_{14}H_{30} \rightarrow C_{10}H_{22} + 2C_2H_4$$

→ Unsaturated hydrocarbons are used to make synthetic polymers.

Alkenes

→ Unsaturated hydrocarbons (C_nH_{2n}) starting at $n = 2$
→ Names are based on the alkanes ('a' changes to 'e')

$CH_2=CH_2$	$CH_3CH=CH_2$	$CH_3CH=CHCH_3$	$C_3H_7CH=CH_2$
ethene	propene	but-2-ene(s)	pent-1-ene
$C_4H_9CH=CH_2$	$C_3H_7CH=CHCH_3$	$C_2H_5CH=CHC_2H_5$	
hex-1-ene	hex-2-ene	hex-3-ene	

Reactions of alkenes ●●●

Alkenes combust to CO_2 and H_2O but they also react by *addition*.

Halogen addition

→ Alkenes rapidly 'decolourize' bromine (brown) in organic solvents: $CH_2=CH_2 + Br_2 \rightarrow CH_2Br-CH_2Br$ or in water: $CH_2=CH_2 + Br_2 + H_2O \rightarrow CH_2Br-CH_2OH + HBr$

The C=C double bond is also rapidly oxidized by warm alkaline aqueous potassium manganate(VII). In this test the purple colour disappears and a brown precipitate is formed in the colourless solution.

Catalytic hydrogenation

→ Hydrogen reduces alkenes to alkanes by adding across the double bond to form a single bond: $CH_2=CH_2 + H_2 \rightarrow CH_3-CH_3$.
→ Transition metals like platinum, palladium and nickel (the cheapest) are used commercially as catalysts.

Mechanism of addition reactions ●●●

→ Reaction of bromine (or hydrogen bromide) with an alkene is an *electrophilic addition*.

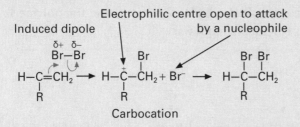

If bromine is in water or NaCl(aq), other nucleophiles can compete with the bromide ion and attack the electrophilic carbon atom. For example, a water molecule (instead of a bromine molecule) may attack the C=C bond and release the nucleophilic OH^- ion.

→ When HBr adds to an unsymmetrical alkene, Br adds to the carbon with the smaller number of H-atoms (Markovnikov's rule).
→ Propene forms 2-bromopropane (*not* 1-bromopropane) in an electrophilic addition reaction with HBr.

Exam question (10 min) answer: page 168

(a) Explain the difference between electrophilic addition and nucleophilic substitution.

(b) Show the mechanism for the chlorination of ethene to form 1,2-dichloroethane and state the conditions for the reaction.

(c) Give the names and structural formulae of *three* possible organic compounds formed when ethene is passed into an aqueous solution of bromine and sodium chloride.

The jargon

−1−, −2−, −3− is the position of double bond.

The jargon

Geometric isomerism arises from restricted rotation about a carbon–carbon double bond, e.g. in but-2-enes:

Steric repulsion   Less steric repulsion

cis-isomer   *trans*-isomer
less stable

Examiner's secrets

If you are asked to give a laboratory test for cyclohexene, you should describe how to do the Br_2 or $KMnO_4$ *and* what to observe for a positive result. Cyclohexene is a liquid but does the same thing as ethene.

The jargon

Markovnikov's rule is $HX + R_2HC=CH_2R \rightarrow R_2HCBr-CH_2R$ because the order of carbonium ion stability is tertiary > secondary > primary or $R_3C^+ > R_2HC^+ > RH_2C^+$.

Hydrocarbons: alkenes and arenes

Saturated hydrocarbons are dull compared to unsaturated compounds like the ethene and benzene. The double bond lets us turn alkenes into alcohols, polymers, etc., by addition reactions. The delocalized benzene ring lets us turn arenes into azo dyes, explosives, etc., by substitution reactions. So how do we make polymers from alkenes?

Manufacture of synthetic polymers from alkenes

→ Ethene polymerizes to poly(ethene) or 'polythene'.
→ Polymerization occurs when a large number of small molecules (monomers) combine to form a large molecule (polymer).
→ Alkenes (and their derivatives) are used to form synthetic polymers.
→ **Low-density poly(ethene)** is formed by heating ethene under pressure with a *trace* of oxygen: $nCH_2=CH_2 \rightarrow (-CH_2-CH_2)_n-$.
→ **High-density poly(ethene)** is made at much lower pressures and temperatures using Ziegler–Natta catalysts (e.g. $Al(C_2H_5)_3$ and $TiCl_4$).

The high-density polymer is tougher and denser than the low-density product because it is more crystalline.

There is a whole range of polymers based on a substituted ethene monomer. The repeat unit can be written as $-CH_2-CHX-$ (where X could be for example H, Cl, CH_3, C_6H_5, etc.) and the polymerization could be represented as follows:

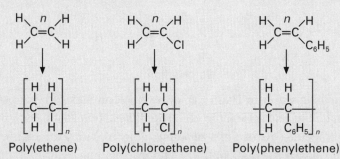

| Poly(ethene) | Poly(chloroethene) | Poly(phenylethene) |

Manufacture of alcohols from alkenes

Ethene is an important industrial feedstock in the large-scale manufacture of a variety of chemicals:

1 direct hydration to ethanol

$$C_2H_4(g) + H_2O(g) \xrightarrow[H_3PO_4 \text{ catalyst}]{70 \text{ atm } 300\,°C} C_2H_5OH(g)$$

2 indirect hydration to ethane-1,2-diol

$$C_2H_4(g) + {}^1\!/_2 O_2(g) \xrightarrow{\text{silver catalyst}} \underset{\text{epoxyethane}}{CH_2-CH_2(g)}$$

$$\underset{O}{CH_2-CH_2}(g) + H_2O(g) \xrightarrow{2 \text{ atm } 150\,°C} CH_2OH-CH_2OH(g)$$

Ethane-1,2-diol is one of the monomers in the production of the polyester 'Terylene'.

The jargon

Greek: *polys* – many.
Poly(ethene), poly(chloroethene), etc. are named using the prefix 'poly' + (monomer name).
In *addition polymerization* monomers combine without the elimination of any other molecules.
In *condensation polymerization* monomers combine with the elimination of other (small) molecules.

Checkpoint 1

Draw a diagram to show the formation of poly(propene).

Watch out!

Poly(2-methylpropenoate) or polymethylmethacrylate = 'perspex' is also an addition polymer (*not* a condensation polymer or polyester).

Don't forget!

The catalytic hydrogenation of alkenes (and unsaturated oils) to alkanes (and saturated fats).

The jargon

Ethane-1,2-diol = ethylene glycol = glycol = antifreeze.

Benzene

Benzene does *not* react like an alkene with three C=C bonds and is 150 kJ mol^{-1} more stable than Kekulé's structure predicts. In fact the 12 atoms are coplanar: the six C-atoms form a regular hexagon and the C–C bond length is between C—C and C=C.

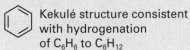 Kekulé structure consistent with hydrogenation of C_6H_6 to C_6H_{12}

 Molecular orbital model with sideways overlap of p-orbitals forming the delocalized π-electrons above and below the ring

→ now represents the delocalized structure of benzene

Watch out!

It's easy to forget the H-atoms attached to the ring.

Reactions of arenes

The π-electrons above and below the ring make benzene open to attack by electrophiles.

→ Arenes (nucleophiles) mainly undergo substitution reactions.
→ These reactions are *electrophilic substitution reactions*.

Nitration of benzene

Benzene reacts with a nitrating mixture of conc. nitric acid in conc. sulphuric acid to form nitrobenzene.

1 Nitronium ions are formed in the nitrating mixture:

$$2H_2SO_4 + HNO_3 \rightarrow NO_2^+ + H_3O^+ + 2HSO_4^-$$

2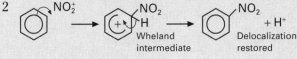

3 $H^+ + HSO_4^- \rightarrow H_2SO_4$

Overall reaction: $C_6H_6 + HNO_3 \rightarrow C_6H_5NO_2 + H_2O$

The jargon

Nitronium ion = nitryl cation = NO_2^+.
Wheland intermediate = cation with four π-electrons shared by five carbon atoms in the benzene ring.

Chlorination of methylbenzene

→ Chlorine reacts with the side chain in UV light and reacts with the benzene ring in the absence of light (and heat) but in the presence of a carrier (e.g. $AlCl_3$).

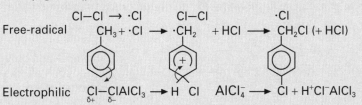

→ Free radical chlorination of the methyl group can replace all three H-atoms to give $C_6H_5CH_2Cl$, $C_6H_5CHCl_2$ and $C_6H_5CCl_3$.

Nitration of the ring can give 2,4,6-trinitrobenzene (TNT).

The jargon

A *side chain* is a methyl group or any other alkyl group attached to the ring.

The jargon

methylbenzene = toluene
TNT = trinitro toluene

Checkpoint 2

Write the structure of TNT and a balanced equation for its formation from methylbenzene.

Exam question (5 min)   answer: page 168

(a) Write an equation for the chlorination of methylbenzene to
    (i) (trichloromethyl)benzene and (ii) 1-chloro-2-methylbenzene.

(b) Describe and explain the mechanism of the reaction between
    methylbenzene and chlorine in the presence of UV light.

Compounds containing halogens

In organic compounds, halogens are joined to carbon by single covalent bonds. The strength of the bond depends on the halogen and on the other atoms joined to the carbon. C—F is strongest and C—I weakest. C_6H_5Cl is unreactive and CH_3COCl very reactive.

The jargon

A *halogenoalkane* has a halogen substituted into an alkane chain.

The jargon

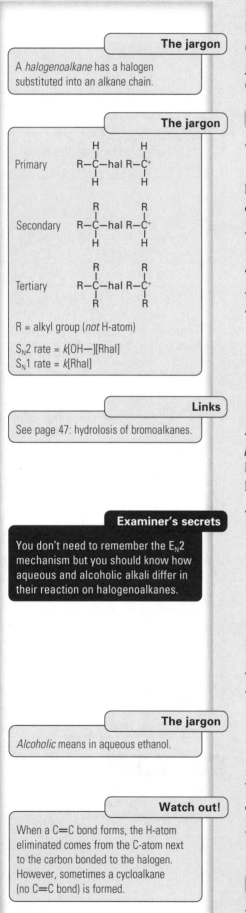

R = alkyl group (*not* H-atom)

S_N2 rate $= k[OH—][Rhal]$
S_N1 rate $= k[Rhal]$

Links

See page 47: hydrolosis of bromoalkanes.

Examiner's secrets

You don't need to remember the E_N2 mechanism but you should know how aqueous and alcoholic alkali differ in their reaction on halogenoalkanes.

The jargon

Alcoholic means in aqueous ethanol.

Watch out!

When a C=C bond forms, the H-atom eliminated comes from the C-atom next to the carbon bonded to the halogen. However, sometimes a cycloalkane (no C=C bond) is formed.

Halogenoalkanes

→ Iodo compounds tend to be the most reactive because the C—I bond is the weakest (relative bond strength is C—Cl>C—Br>C—I).

Halogenoalkanes are attacked by nucleophiles, undergoing substitution or elimination reactions depending upon the reagent and conditions.

→ The electrophilic centre is the carbon atom attached to the halogen.

Alkaline hydrolysis

→ Nucleophilic *substitution* with *aqueous* sodium hydroxide
→ S_N2 reaction mechanism for *primary* halogenoalkanes

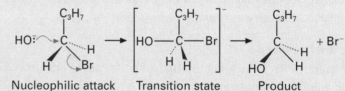

Nucleophilic attack   Transition state   Product

A similar S_N2 mechanism applies to the nucleophilic substitution of *primary* halogenoalkanes by CN^- (or NH_3) to form nitriles (or amines). Kinetics experiments indicate that tertiary (and some secondary) halogenoalkanes react by a different mechanism.

→ S_N1 reaction mechanism for *tertiary* halogenoalkanes

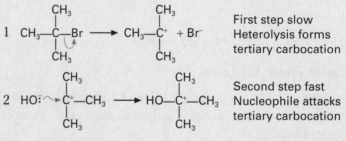

1   First step slow — Heterolysis forms tertiary carbocation

2   Second step fast — Nucleophile attacks tertiary carbocation

→ Nucleophilic *elimination* with *alcoholic* potassium hydroxide
→ E_N2 reaction mechanism for *primary* halogenoalkanes

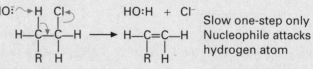

Slow one-step only — Nucleophile attacks hydrogen atom

You can think of elimination (favoured by ethanol as solvent) competing with substitution (favoured by water as solvent).

→ Elimination reactions occur with tertiary halogenoalkanes more readily than with primary halogenoalkanes.

Aromatic halogen compounds ●●●

→ Halogens attached to the benzene ring are not reactive because their p-electrons become part of the ring's delocalized π-system.
→ Halogen attached to the side chain behaves like halogenoalkane.

Acyl chlorides

The two compounds you meet at A-level are ethanoyl chloride (CH_3COCl) and benzoyl chloride (C_6H_5COCl).

→ Acyl chlorides are very reactive.

Electronegative oxygen and chlorine make the attached carbon strongly electrophilic and readily attacked by nucleophiles

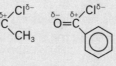

→ Acyl chlorides fume in moist air: $RCOCl + H_2O \rightarrow RCO_2H + HCl$.
→ $CH_3COCl(l)$ and $C_6H_5COCl(l)$ are powerful acylating agents used, for example, to make esters and amides.

Phenol (unlike alcohols) will not react directly with carboxylic acids to form esters.

→ Phenol will react with acyl chlorides to form esters.

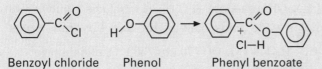

Benzoyl chloride   Phenol   Phenyl benzoate

Uses of halogen compounds

Their commercial and industrial importance depends upon their ability to take part in a variety of nucleophilic substitution reactions.

→ As intermediates in synthetic organic chemistry: e.g. extending the carbon chain $ROH \rightarrow RBr \rightarrow RCN \rightarrow RCO_2H \rightarrow RCH_2OH$.
→ Chloroethane was used to make tetraethyl lead (main antiknock agent in leaded petrol).
→ 1,2-dibromoethane was used as a lead scavenger in the fuel.
→ Chlorofluorocarbons have been used as refrigerants and as propellants in aerosols.
→ Chloroethene is the monomer for making PVC.
→ Chlorinated hydrocarbons such as tetrachloromethane, 1,1,1-trichlorethane and trichloromethane are good solvents.
→ Chlorobenzene is used in the manufacture of DDT.

Environmental concerns

The EU has banned CFCs to avert further damage to the ozone layer. The use of the insecticide DDT against malaria-carrying mosquitoes is restricted because it dissolves in animal fat and accumulates in our food chain. Vinyl chloride is very toxic and disposal of PVC by burning produces highly toxic chlorinated dioxins. Chlorinated solvents can irreversibly damage internal organs if they are ingested or inhaled. Some, like CCl_4, are dangerously carcinogenic.

Exam question (3 min)   answer: page 169

Draw and explain the mechanism for the substitution reaction of potassium cyanide with 1-bromopropane.

The jargon

Acyl group = R—CO— = $R\!\!>\!\!C\!=\!O$
R = alkyl or aryl
R ≠ H, OH, NH$_2$
In *acylation* RCO is substituted for H in —OH, —NH$_2$, etc.
Substituting CH_3CO— is *ethanoylation*.

Links

See page 153: formation of esters by carboxylic acids.

Checkpoint

Draw structures of the products for ethanoyl chloride reacting with
(i) ethanol
(ii) phenol
(iii) phenylamine
(iv) ethylamine

The jargon

Chloroethene or *vinyl chloride* = CH_2=CHCl
Trichloromethane or *chloroform* = $CHCl_3$
1,1,1-trichlorethane = CH_3CCl_3
1,1,1-trichloro-2,2-di(4-chlorophenyl)ethane = DDT =

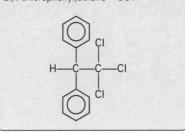

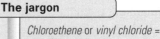

Compounds containing oxygen: alcohols, phenols,

For A-level you study alcohols and phenol, aldehydes and ketones, carboxylic acids and their derivatives.

Alcohols

→ Hydroxyl (—OH) is the functional group of alcohols.
→ OH attached to a benzene ring is the functional group of phenols.

We class alcohols as *primary*, *secondary* or *tertiary* according to the number of H-atoms attached to the carbon atom of the C—OH group.

Aliphatic alcohols

CH_3OH	methanol	$C_2H_5CH_2OH$	propan-1-ol
C_2H_5OH	ethanol	$CH_3CH(OH)CH_3$	propan-2-ol
$C_3H_7CH_2OH$	butan-1-ol	$C_2H_5CH(OH)CH_3$	butan-2-ol

Formation of alcohols

→ In general alcohols are formed by hydrolysis of halogenoalkanes with the bromo and iodo compounds giving the best yields:
$C_2H_5I + NaOH(aqueous) \rightarrow C_2H_5OH + NaI$ [heat under reflux]
→ Ethanol is made industrially by the direct hydration of ethene

$C_2H_4 + H_2O \rightarrow C_2H_5OH$ (70 atm, 300 °C, phosphoric acid catalyst)

and by fermentation of carbohydrates

$C_6H_6O_6 \rightarrow 2C_2H_5OH + 2CO_2$ (catalysed by enzymes from yeast)

Reactions of alcohols

→ Alcohols react with metallic sodium to give hydrogen gas:

$$C_2H_5OH + Na \rightarrow C_2H_5O^-Na^+ + \tfrac{1}{2}H_2$$

→ Alcohols are oxidized by acidified potassium dichromate(VI)

$C_2H_5OH \rightarrow CH_3CHO$ (continuously distilled from oxidant)
$C_2H_5OH \rightarrow CH_3CHO \rightarrow CH_3COOH$ (refluxed with oxidant)

→ Alcohols form esters when refluxed with carboxylic acids and a mineral acid catalyst:

→ Alcohols react with PCl_5 to give fumes of hydrogen chloride:

$C_2H_5OH + PCl_5 \rightarrow C_2H_5Cl + POCl_3 + HCl$ (test for OH groups)

Uses of alcohols

→ Methanol: manufacture of 'Perspex', methanoic and ethanoic acids
→ Ethanol: fuel, solvent and ingredient in wines, beers and spirits
→ Propan-2-ol: solvent and manufacture of propanone, phenol, etc.
→ Ethan-1,2-diol: antifreeze and manufacture of polyesters

Phenols

→ C_6H_5OH is the simplest phenol and a slightly water-soluble weak acid (carbolic acid) – conjugate base is $C_6H_5O^-$ (phenoxide ion):
$C_6H_5OH + H_2O \rightleftharpoons C_6H_5O^- + H_3O^+$; $K_a = 1.3 \times 10^{-10}$ mol dm^{-3}

aldehydes and ketones

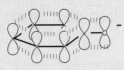 phenoxide ion stabilized by sideways overlapping p-orbitals of oxygen becoming part of the ring's delocalized π-system

→ Phenol dissolves in aqueous sodium hydroxide to form a solution of sodium phenoxide: $C_6H_5OH + NaOH \rightarrow C_6H_5ONa + H_2O$.

→ Phenol is weaker than carbonic acid, so carbon dioxide displaces it from the salt: $C_6H_5ONa + CO_2 + H_2O \rightarrow C_6H_5OH + NaHCO_3$.

Differences between phenol and ethanol

→ Aqueous phenol is acidic.

→ Phenol does not react with hydrogen halides, PBr_3 or PI_3.

→ Phenol resists oxidation but strong oxidants produce a complex mixture of aliphatic and aromatic products.

Tests for phenol

→ Bromine water gives a white precipitate (2,4,6-tribromophenol).

→ Aqueous iron(III) chloride gives characteristic violet colours with most phenolic compounds.

Uses of phenols

→ TCP (trichlorophenol) is a mixture of chlorinated phenols

→ manufacture of phenolic (Bakelite) and epoxy resins

→ manufacture of cyclohexanol for producing nylon

Aldehydes and ketones ●●●

For A-level, ethanal, CH_3CHO, and propanone, $(CH_3)_2CO$, are the most important carbonyl compounds. Methanal (formaldehyde), HCHO, is the simplest aldehyde but it has some atypical properties.

Aldehydes

→ Laboratory preparation by oxidation of alcohols controlled to avoid further oxidation to carboxylic acids

→ Catalytic oxidation of ethanol vapour over copper catalyst

→ Commercial manufacture of ethanal from

1   ethene and steam using palladium(II) chloride catalyst

2   ethanol by air oxidation and dehydrogenation with silver catalyst: $CH_3CH_2OH + \frac{1}{2}O_2 \rightarrow CH_3CHO + H_2O$ (exothermic) gives heat to $CH_3CH_2OH \rightarrow CH_3CHO + H_2$ (endothermic).

Exam question (10 min)   answer: page 169

(a) State how and under what conditions ethanol reacts with acidified dichromate.

(b) Explain how (i) ethanol is manufactured from ethene by direct hydration, (ii) glycerol is formed by the alkaline hydrolysis of oils and fats.

The jargon

These are compounds in which the —OH group is attached directly to the benzene ring. The simplest phenol is phenol itself, C_6H_5OH.

Checkpoint 2

Write an ionic equation for the reaction of phenol with NaOH(aq).

The jargon

2,4,6-tribromophenol = [structure: benzene ring with OH at top, Br at positions 2, 4, 6]

Watch out!

Iron(III) chloride test works best in 'neutral' solution: too acidic – too little $C_6H_5O^-$ ion; too alkaline – too little Fe^{3+}.

The jargon

A *carbonyl group* is >C=O. *Carbonyl compounds* are

$R_{\delta+}^{}C=\overset{\delta-}{O}$ with H $R_{\delta+}^{}C=\overset{\delta-}{O}$ with R

aldehydes   and   ketones (R ≠ H)

Watch out!

This catalytic oxidation of $C_2H_5OH \rightarrow CH_3CHO$ needs a fume cupboard.

copper wire kept hot by exothermic reactions

warm ethanol

Compounds containing oxygen: aldehydes, ketones

CHO

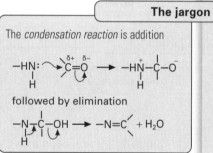
The key to the chemistry of this group is the polarity of the >C=O group.

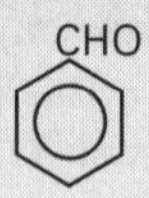

Reactions of aldehydes

→ Aldehydes (but *not* ketones) are reducing agents identified by the following popular tests:

Test	*Positive observation*
→ Warm with Fehlings solution: complex Cu(II) ions in alkali	→ Colour changes from blue to green and finally get red precipitate Cu_2O
→ Warm with Tollens' reagent: complex Ag(I) ions in alkali	→ Silver mirror forms on inside of test tube or black precipitate deposited
→ Warm with acidified aqueous potassium dichromate	→ Colour of solution turns from orange to green

Reactions of ketones

→ Ketones are *not* readily oxidized.

Methyl ketones (and compounds easily oxidized to a methyl ketone) will produce a yellow precipitate of triiodomethane (iodoform) when warmed with aqueous sodium hydroxide and iodine (or aqueous sodium chlorate(I) and aqueous potassium iodide). The *iodoform reaction* is a test for CH_3CO- group or a group, like $CH_3CH(OH)-$, easily oxidized to it.

Reactions of aldehydes and ketones

→ The carbonyl group >C=O is attacked by nucleophiles.

Condensation reactions

→ Carbonyl compounds react with 2,4-dinitrophenylhydrazine to form yellow/orange precipitates of 2,4-dinitrophenylhydrazones:

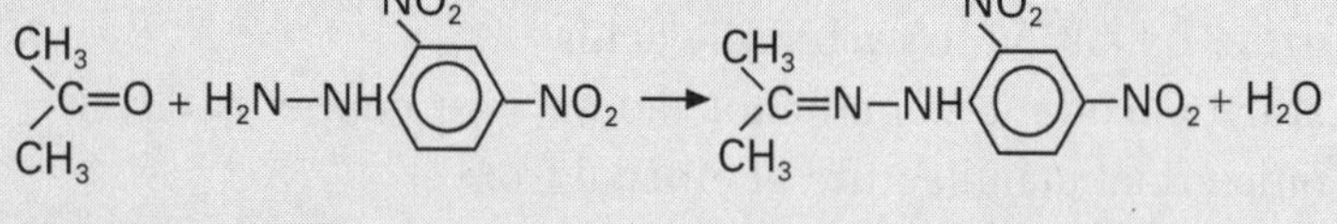

Addition reactions

→ HCN and HSO_3^- add to >C=O to give >C(OH)CN and >C(OH)SO_3^-:

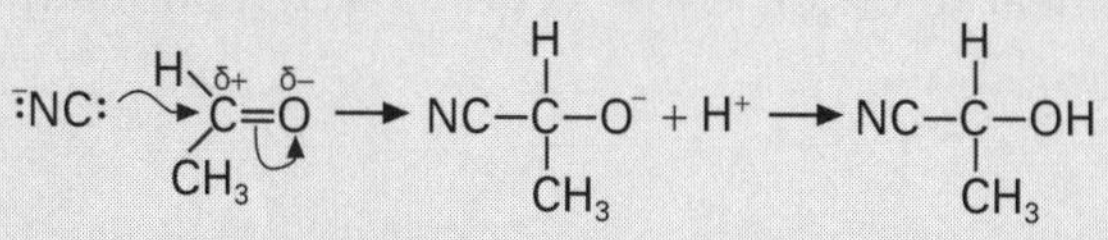

aqueous sodium or potassium cyanide is followed by excess mineral acid

Reduction

→ Sodium tetrahydridoborate(III), $NaBH_4$, reduces aldehydes to *primary* alcohols and ketones to *secondary* alcohols:
ethanal CH_3CHO [+ 2H] → CH_3CH_2OH ethanol
propanone $(CH_3)_2CO$ [+ 2H] → $(CH_3)_2CHOH$ propan-2-ol

and carboxylic acids

Carboxylic acids ●●○

Methanoic acid and ethanoic acid are the first two homologues of the straight-chain aliphatic monobasic weak carboxylic acids.

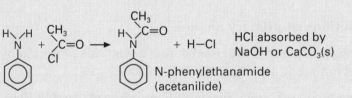

Ethanoate ion stabilized by delocalization

→ Ethanoic acid is stronger than carbonic acid and (unlike phenol) liberates CO_2 from carbonates and hydrogencarbonates:
$$CH_3CO_2H + NaHCO_3 \rightarrow CH_3CO_2Na + H_2CO_3 \rightarrow H_2O + CO_2$$

Aromatic carboxylic acids like benzoic acid and 2-hydroxybenzoic acid dissolve in hot water but less well than aliphatic acids in cold water.

Reactions of carboxylic acids

→ Lithium tetrahydridoaluminate(III), $LiAlH_4$, in dry ether reduces the carboxyl group to primary alcohol: $RCOOH \rightarrow RCH_2OH$
→ PCl_5 (or $SOCl_2$) attack the OH in the carboxyl group to form acyl chloride: $RCOOH + PCl_5 \rightarrow RCOCl + POCl_3 + HCl$
→ Refluxing with alcohols and a mineral acid catalyst (H_2SO_4 or HCl) esterifies the carboxyl group to form sweet smelling neutral compounds: $CH_3COOH + C_2H_5OH \rightleftharpoons CH_3COOC_2H_5 + H_2O$ ethyl ethanoate (an ester)
→ Decarboxylation (heating with soda lime: CaO(s) + NaOH(aq)) forms hydrocarbons: $RCO_2H \rightarrow RH + CO_2$ ($\rightarrow CaCO_3$; Na_2CO_3)

Carboxylic acid derivatives ●●●

Acyl chlorides and acid anhydrides

→ Reactive compounds readily hydrolysed by water back to carboxylic acids: $CH_3COCl + H_2O \rightarrow CH_3COOH + HCl$
→ React with (acylate) OH and NH_2 groups replacing an H with an acyl group (RCO): $CH_3COCl + HOCH_3 \rightarrow CH_3COOCH_3 + HCl$

HCl absorbed by NaOH or CaCO₃(s)

N-phenylethanamide (acetanilide)

Amides and esters ●●●

→ On refluxing with NaOH(aq) amides give ammonia and esters give the alcohol and sodium salt of the carboxylic acid:
ethanamide: $CH_3CONH_2 + NaOH \rightarrow CH_3CO_2Na + NH_3$
ethyl methanoate: $HCO_2C_2H_5 + NaOH \rightarrow HCO_2Na + C_2H_5OH$

Exam question (6 min)   answer: page 169

(a) Write an equation for ethanoyl chloride reacting with 2-hydroxybenzoic acid. Draw the structure of the organic product. (b) How could you prepare methyl 2-hydroxbenzoate (oil of wintergreen) from 2-hydroxybenzoic acid.

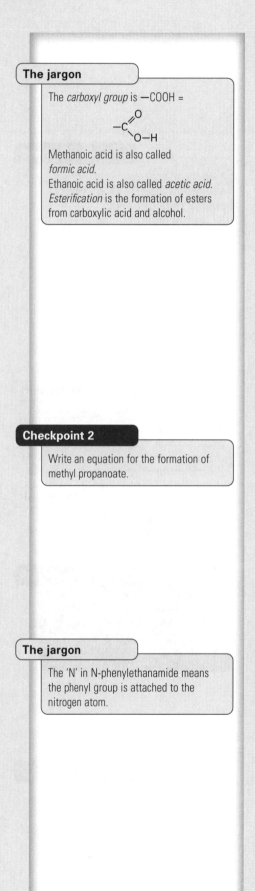

The jargon

The *carboxyl group* is —COOH =

$-C\overset{O}{\underset{O-H}{}}$

Methanoic acid is also called *formic acid*.
Ethanoic acid is also called *acetic acid*.
Esterification is the formation of esters from carboxylic acid and alcohol.

Checkpoint 2

Write an equation for the formation of methyl propanoate.

The jargon

The 'N' in N-phenylethanamide means the phenyl group is attached to the nitrogen atom.

Compounds containing nitrogen: amines and

All A-level syllabuses refer to amines, amides, polyamides, amino acids and proteins. The key to your study of these and other related substances is the amino group, $-NH_2$, and the amide (peptide) link $-CO-NH-$.

Checkpoint 1

Write the structural formula of the following and identify each as a primary, secondary or tertiary amine:
(i)   2-aminobutane
(ii) 2-methylphenylamine
(iii) ethyldimethylamine
(iv) N:N-dimethylphenylamine

Amines

Amines are derivatives of ammonia classified by the number of carbons attached to the nitrogen atom:

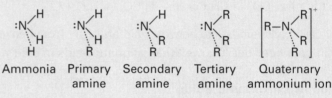

Ammonia | Primary amine | Secondary amine | Tertiary amine | Quaternary ammonium ion

→ Ammonia and amines are *weak* bases (proton acceptors) and nucleophiles because of the lone electron pair on the N-atom:
$NH_3 + H_2O \rightleftharpoons NH_4^+ + OH^-$ and $RNH_2 + H_2O \rightleftharpoons RHN_3^+ + OH^-$
→ Ethylamine is slightly stronger base and phenylamine a slightly weaker base than aqueous ammonia.

Formation of primary amines

→ Nucleophilic substitution reactions of ammonia with halogenoalkanes: $NH_3 + Rhal \rightarrow RNH_2 + HX (\rightarrow NH_4X)$
→ Reduction of nitriles and amides with tetrahydridoaluminate(III):
$RCN [+4H] \rightarrow RCH_2NH_2$; $RCONH_2 [+4H] \rightarrow RCH_2NH_2 + H_2O$
→ Reduction of aromatic nitro compounds, e.g. nitrobenzene
$C_6H_5NO_2[+6H] \rightarrow C_6H_5NH_2 + 2H_2O$ to phenylamine.

Watch out!

When ammonia and halogenoalkanes are heated in sealed tubes the reaction produces a mixture of *p*-, *sec*- and *tert*-amines.

The jargon

tetrahydridoaluminate(III) = $LiAlH_4$ = lithium aluminium hydride.

With nitrobenzene you could use tin and concentrated hydrochloric acid as the reducing agent but the excess acid would form a salt with phenylamine: $C_6H_5NH_2 + HCl \rightarrow C_6H_5NH_3^+ + Cl^-$. So you must add excess NaOH to free the amine and then extract it from the product mixture by *steam distillation*.

Examiner's secrets

There are often questions about the preparation of phenylamine from nitrobenzene because the process involves some important chemical principles.

Reactions of primary amines

→ NH_2 group reacts with (and during a synthesis protected by) acyl chlorides: $RNH_2 + CH_3COCl \rightarrow RNHCOCH_3 + HCl$

The primary NH_2 group reacts with nitrous acid (from HCl(aq) and sodium nitrite, $NaNO_2$) to form unstable diazonium ions:

$$RNH_{3(aq)}^+ + HNO_2(aq) \rightarrow RN_2^+(aq) + 2H_2O(l)$$

Checkpoint 2

Write an equation to suggest how urea (carbamide), $CO(NH_2)_2$, might react with nitrous acid.

With *aliphatic* amines, even in ice-cold conditions, nitrogen gas is liberated almost quantitatively, $RN_2^+(aq) \rightarrow R^+(aq) + N_2$, and the very unstable electrophilic carbocation combines with various nucleophiles giving a mixture of organic products including RCl, RNO_2, ROH, etc. If conditions are *not* ice-cold, a similar reaction occurs with *aromatic* primary amines: $C_6H_5NH_2 + HNO_2(aq) \rightarrow C_6H_5OH + H_2O + N_2$.

→ If conditions are ice-cold, the *diazotization* of an *aromatic primary amine* forms a stable *aqueous aromatic diazonium cation*.

amino acids

Azo-dyes

→ Diazotization is the formation of stable aqueous diazonium cations by reaction of a primary aromatic amine with nitrous acid at 0–5 °C:

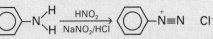

Phenylamine in HCl(aq)   Benzenediazonium chloride

→ Benzenediazonium chloride is too unstable to be isolated but we use its solution in coupling reactions with phenols or aromatic amines to produce azo-dyes:

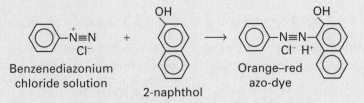

Benzenediazonium   2-naphthol   Orange–red
chloride solution   azo-dye

Because diazonium compounds couple with phenols and amines it is very important during the preparation of benzenediazonium chloride
1 to keep the temperature below 10 °C so no phenol is formed
2 to diazotize all the phenylamine so none is left to couple
→ Bromine water reacts with a solution of phenylamine in HCl(aq) to give a whitish precipitate of 2,4,6-tribromophenylamine:

$$C_6H_5NH_2(aq) + 3Br_2(aq) \rightarrow C_6H_2Br_3NH_2(s) + 3HBr(aq)$$

Amino acids ●●●

→ Amino acids contain an amino group and a carboxyl group:

2-aminopropanoic acid NH_2 NH_2 3-aminopropanoic acid
α-aminopropanoic acid CH_3CHCO_2H $CH_2CH_2CO_2H$ β-aminopropanoic acid

α-amino acids

→ About 20 α-amino acids (2-aminoalkanoic acids) occur naturally and produce proteins essential to life.
→ The R group may be one of about 20 different groups:

	neutral		acidic	basic
side-group =				
—R =	—H	—CH₃	—CH₂CO₂H	—CH₂(CH₂)₃NH₂
name =	glycine	alanine	glutamic acid	lysine
code =	gly	ala	glu	lys

Glycine, $CH_2NH_2CO_2H$, is the simplest α-amino acid and is typical in forming a solid with a crystal lattice of *zwitterions*: $^+NH_3CH_2CO_2^-$.

→ In alkali (pH > 7) the zwitterion becomes $NH_2CH_2CO_2^-$
In acid (pH < 7) the zwitterion becomes $^+NH_3CH_2CO_2H$

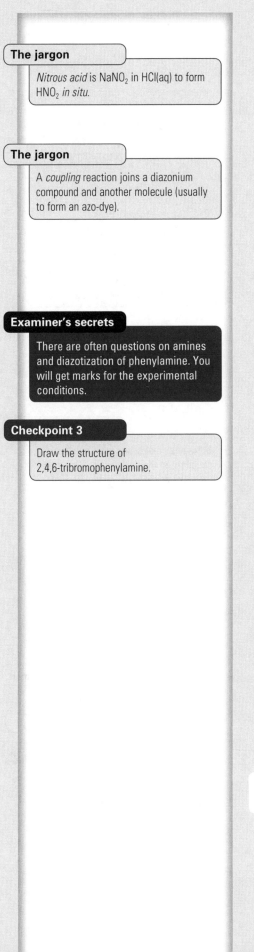

The jargon

Nitrous acid is $NaNO_2$ in HCl(aq) to form HNO_2 *in situ*.

The jargon

A *coupling* reaction joins a diazonium compound and another molecule (usually to form an azo-dye).

Examiner's secrets

There are often questions on amines and diazotization of phenylamine. You will get marks for the experimental conditions.

Checkpoint 3

Draw the structure of 2,4,6-tribromophenylamine.

Exam question (5 min)   answer: page 170

Suggest why, during diazotization (a) phenylamine is in HCl(aq), (b) coupling agents like 2-naphthol are in alkaline solution, and (c) benzenediazonium ions are more stable than ethyldiazonium ions.

Compounds containing nitrogen: proteins and

Polyamides occur naturally as proteins. Proteins are built from long chains of amino acids linked by peptide bonds. Shorter chains are called polypeptides. Nylons are synthetic polyamides.

Proteins ●●●

Optical isomers

All the α-amino acids *except* glycine have chiral molecules with four different atoms or groups attached to a carbon atom, so they can show *optical isomerism*.

→ All naturally occurring amino acids from protein are L-forms.

An amino group in one molecule and a carboxyl group in another molecule can *in principle* lose a water to form an amide link:

$$NH_2-\overset{H}{\underset{R}{C}}-CO_2H + NH_2-\overset{H}{\underset{R}{C}}-CO_2H \xrightarrow{-H_2O} NH_2-\overset{H}{\underset{R}{C}}-\overset{O}{C}-\overset{H}{N}-\overset{H}{\underset{R}{C}}-CO_2H = \text{a dipeptide}$$

M_r of proteins ranges from 5×10^3 to 4×10^7. Polypeptides like oxytocin (causes uterine contraction in childbirth) are proteins with small relative molecular masses.

→ Proteins are natural polyamides with only one —CHR— between each amide (peptide) linkage:

$$NH_2-\overset{H}{\underset{R_1}{C}}-\overset{O}{C}-\overset{H}{N}-\overset{H}{\underset{R_2}{C}}-\overset{O}{C}-\overset{H}{N}-\overset{H}{\underset{R_3}{C}}-\overset{O}{C}-\overset{H}{N}-\overset{H}{\underset{R_4}{C}}-\overset{O}{C}-OH$$

→ The R group may be any one of about 20 different groups in any order along the protein chain.
→ The *primary structure* of a protein is the sequence of amino acids in the chain, often shown by codes: GlyAlaLysGluGluSerMet. . . .
→ S—S bonds, hydrogen bonding and electrostatic attractions involving side groups cause chains to cross-link, coil into a helix and fold into pleats to produce the *secondary structure* of proteins.
→ The *tertiary structure* is the overall three-dimensional shape of (cross-linked, coiled and pleated) protein molecules.

Enzymes, antibodies, haemoglobin, casein, albumin and insulin are globular proteins soluble in water. Keratin and collagen (in hair and muscle) are fibrous proteins insoluble in water.

Proteins and chromatography

→ Chromatography separates and identifies components in a mixture.

We can hydrolyse a protein with HCl(aq) to a mixture of amino acids

$$\ldots -CHR-CO-\!\!-NH-CHR'- \ldots H_2O$$
$$\rightarrow \ldots -CHR-CO_2H \quad NH_2-CHR'- \ldots$$

and use *paper chromatography* to separate the mixture. We spray the paper with ninhydrin. On drying and warming, purple spots appear and the amino acids are identified by comparing their R_f value with data book values.

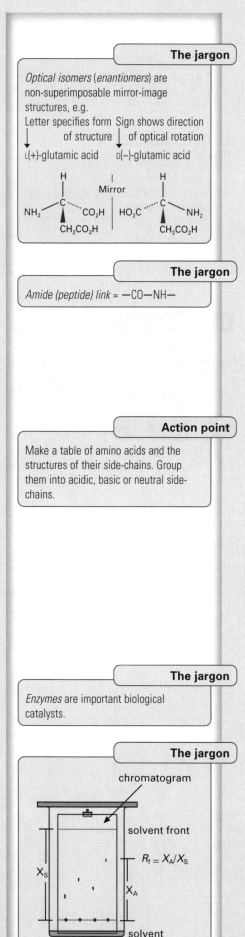

The jargon

Optical isomers (enantiomers) are non-superimposable mirror-image structures, e.g.

Letter specifies form of structure — L(+)-glutamic acid

Sign shows direction of optical rotation — D(–)-glutamic acid

Mirror

The jargon

Amide (peptide) link = —CO—NH—

Action point

Make a table of amino acids and the structures of their side-chains. Group them into acidic, basic or neutral side-chains.

The jargon

Enzymes are important biological catalysts.

The jargon

chromatogram

solvent front

$R_f = X_A/X_S$

X_S

X_A

solvent

polyamides

Amides and polyamides ●●●

Amides

Amides have the general formula $RCONH_2$ and are derived from carboxylic acids RCOOH by replacing the OH with NH_2:

$$CH_3COOH + NH_3 \rightarrow CH_3COONH_4 \xrightarrow{\text{heat}} CH_3CONH_2 + H_2O$$
$$CH_3COCl + 2NH_3 \rightarrow CH_3CONH_2 + NH_4Cl$$

Polyamides

Proteins are *natural* polyamides and nylons are *synthetic* polyamides with repeating units joined by the peptide or amide link which can be represented by

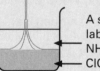, $-CO-NH-$ or $-CONH-$

In different nylons the group between each link will differ:

nylon-6–10
$$NH-(CH_2)_6-NH-CO-(CH_2)_8-CO-NH-(CH_2)_6-NH-CO-(CH_2)_8-CO-$$
nylon-6–8
$$NH-(CH_2)_6-NH-CO-(CH_2)_6-CO-NH-(CH_2)_6-NH-CO-(CH_2)_6-CO-$$
nylon-6
$$NH-(CH_2)_5-CO-NH-(CH_2)_5-CO-NH-(CH_2)_5-CO-$$

A sample of nylon-6–10 can be prepared in the laboratory at the interface of a solution of $NH_2-(CH_2)_6-NH_2$ in water and $ClCO-(CH_2)_8-COCl$ in tetrachloromethane.

Nylon-6 is made industrially from a single monomer called caprolactam. Under suitable conditions this cyclic amide splits open and polymerizes.

caprolactam

The polymer is spun into fibres that have elasticity and high tensile strength because the long-chain molecules can coil and stretch.

Exam question (12 min)   answer: page 170

(a) Outline the preparation of phenylamine from nitrobenzene.

(b) State how phenylamine reacts with nitrous acid.

(c) Draw the isomers of 2-amino propanoic acid.

(d) State the effect of adding aqueous sodium hydroxide to (i) the ammonium salt of a carboxylic acid; (ii) a primary amine; (iii) an amide of a carboxylic acid.

(e) State and explain what you would expect to happen if nylon is refluxed with aqueous hydrochloric acid.

Checkpoint 1

Suggest an equation for the formation of ethanamide from ammonia and ethanoic anhydride.

Examiner's secrets

Make sure you know
→ the difference between amides and amines
→ an example of a primary, secondary and tertiary amine
→ the relationship between α-amino acids and proteins
→ an example of an important polyamide
→ the differences between proteins and polyamides

Checkpoint 2

Write an equation to show the reaction of $ClCO-(CH_2)_8-COCl$ with $NH_2-(CH_2)_6-NH_2$ to show the formation of nylon-6–10.

Examiner's secrets

Remember that marks are given for the reagents and reaction conditions used.

Instrumental techniques: UV, visible and

Checkpoint 1

Draw and label a line to represent the electromagnetic spectrum. Include radiowaves, microwaves, X-rays, ultraviolet, infrared and visible radiation in order of decreasing frequency.

The jargon

A *chromaphore* is an ion, molecule or group absorbing UV or visible light; e.g. $-N=N-$ absorbs blue so azo-dyes are often orange-red.

Checkpoint 2

(a) At which end of the visible spectrum (red–orange–yellow–green–blue–indigo–violet) is the maximum absorbance of phenolphthalein in basic solution?
(b) Sketch a possible absorption spectrum in the visible region for chlorophyll.

Chemists have combined computer technology and spectroscopy into very powerful tools for analysing and identifying compounds. You need to understand the principles and be able to interpret simple simple spectra.

Ultraviolet and visible spectroscopy

We see colours because (electronic transitions in) compounds absorb light of different wavelengths from the visible region of the electromagnetic spectrum. Electrons move between d-orbitals in d-block compounds and within conjugated systems of π-electrons in organic *chromophoric* compounds. Our eyes cannot see UV light but many compounds absorb light in the UV region.

Split-beam UV/visible spectroscopy

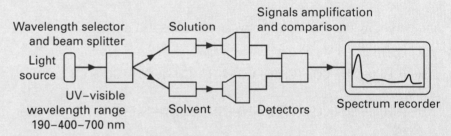

The instrument automatically compares the absorptions by the solution and pure solvent (both in identical glass or quartz cells) to get the absorbance by the solute only. The concentration of the solute in the solution is related to the absorbance by the Beer–Lambert law.

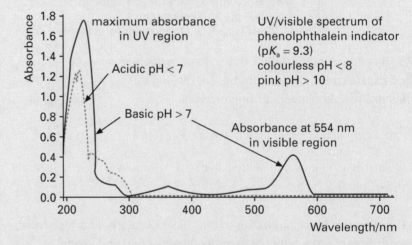

→ UV and visible spectroscopy is used as an analytical tool to study d-block metal ions and complexes, to determine structure and to study the kinetics of chemical reactions.

Infrared spectroscopy

→ Molecules can absorb IR radiation as *vibrational* energy that excites the natural bending and stretching of their bonds:

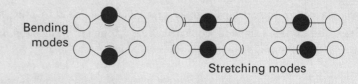

Bending modes

Stretching modes

IR spectroscopy

Infrared spectrometers are similar to UV/visible instruments but the sample must be in a cell made of crystalline sodium chloride because (unlike glass which it resembles) the salt does *not* absorb IR radiation. The instrument automatically

1 varies the IR frequency beamed into the sample
2 detects the radiation transmitted and
3 records the results as a spectrum of transmission against wavenumber

The absorption spectrum is like a molecular 'fingerprint' because the different functional groups in a molecule have different characteristic absorption frequencies. However, the spectra are complicated because these characteristic frequencies may change with the position of the functional group in the molecule.

→ Advantages of IR spectroscopy – we only need a very small sample for a good spectrum and we get results quickly.

Infrared spectrum of hexan-2-one

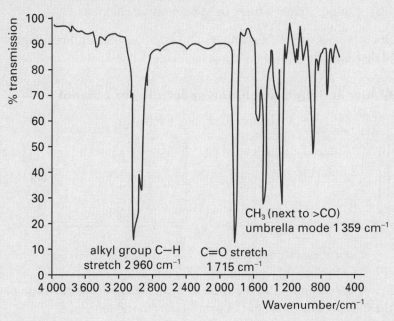

Examiner's secrets

You will always be given these frequencies in an exam. You won't have to remember them.

The jargon

For UV and visible spectroscopy we use wavelength in nm.
For IR spectroscopy we use wavenumber in cm⁻¹.
Wavenumber = 1/wavelength.

Checkpoint 3

Draw a diagram to describe the CH_3 umbrella mode of vibration.

Watch out!

[iodine]/10^{-4} × mol dm⁻³ = 5.40 means the iodine concentration is 0.000 540 mol dm⁻³

The jargon

PABA (para-aminobenzoic acid) = 4-aminobenzoic acid.

Exam questions (6 min, 10 min)   answers: page 171

1 A spectrometer was used to measure the colour of aqueous iodine and follow the change in concentration with time in a reaction mixture. Use the data to plot a suitable graph and find the order of reaction with respect to the iodine.

[iodine]/10^{-4} mol dm⁻³	5.40	4.92	4.28	3.64	2.84	2.20
time/min	0.0	3.0	7.0	11.0	16.0	20.0

2 Discuss briefly how UV and IR spectroscopy might be used (a) to detect a trace impurity in a sample of PABA (an organic compound used in suntan creams) and (b) to identify the gases in a sample from a faulty car exhaust system.

Instrumental techniques: NMR and mass spectrometry

Mass spectrometers and NMR spectrometers are powerful tools for determining the structure of organic compounds. Both kinds of instrument use magnetic fields but in different ways.

NMR spectroscopy ●●●

The nuclear magnetic resonance instrument

1 surrounds the sample with an extremely powerful magnetic field
2 passes radiofrequency waves (from an oscillator) into the sample
3 scans by varying magnetic field strength *or* radiowave frequency
4 detects and records the results as an absorption spectrum

The technique works with the nuclei of ^1H, ^{13}C, ^{19}F and ^{31}P because they have an odd number of protons (or neutrons). A 14.1 T (Tesla) magnetic field strength needs a radio frequency of 60 MHz for ^1H. The frequency of the radiowaves absorbed depends on the environment of the nucleus. Hydrogen atoms in an organic compounds usually have different environments. Consequently, NMR spectroscopy is now one of the techniques most widely used by organic chemists.

→ Doctors now use NMR scanners on people as part of their diagnosis of diseases.

NMR low and high resolution spectrum for ethanol

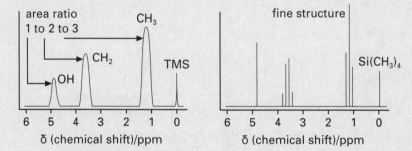

→ In low resolution NMR, the ratio of peak areas tells us the ratio of the numbers of each kind of proton (^1H) in the molecule.

We expect three integrated peaks in ethanol (there are three types of ^1H) with the areas in the ratio $3 : 2 : 1$ (in CH_3, CH_2, and OH).

→ Tetramethylsilane (TMS) is the standard to set the scale because its 12 H-atoms are equivalent and it gives only one peak.

We mix a small amount of TMS with the sample and the instrument generates a chemical shift scale. The chemical shift depends on the particular environment of the proton.

→ In high resolution NMR, peaks split when adjacent carbon atoms have H-atoms attached that are no equivalent.

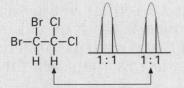 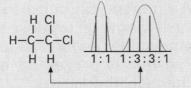

low res. peak for H-atom
splits into two high res.
peaks of equal intensity

low res. peak for H-atom
splits into four high res.
peaks of different intensity

If there are n equivalent hydrogen atoms on the C-atom next to the carbon carrying the ^{1}H, then the low resolution peak for that ^{1}H will split into $n + 1$ high resolution peaks whose relative intensities can be calculated form Pascal's triangle.

n	Relative intensities of peaks
1	1 1
2	1 2 1
3	1 3 3 1
4	1 4 6 4 1

Mass spectrometry ●●●

→ Instrument ionizes (and fragments) gas molecules into a beam of *positive* ions to be scanned by a varying magnetic field.
→ Signal strength (amplitude) of detected current is proportional to the abundance of ions of a given mass/charge ratio.

If some molecules do not fragment, the spectrum may show a peak for the molecular ion and give us a value for the relative molecular mass.

Incomplete mass spectrum of ethanol

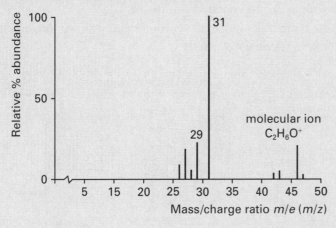

molecular ion $C_2H_6O^+$

Checkpoint

Suggest why mass spectrometers must be kept at a very low pressure.

The jargon

The *mass/charge ratio* is the mass of the ion divided by its positive charge.
A *molecular ion* is a whole (unfragmented) molecule that has lost one electron.

Watch out!

Naturally occurring isotopes of C, H and O complicate high-resolution mass spectra and have to be taken into account.

Exam questions (6 min, 3 min)   answers: page 171

1   For the mass spectrum of ethanol, suggest and explain (a) what causes the peaks at $m/e = 31$ and 29, (b) two possible peaks not shown, (c) why there is a very small peak at $m/e = 47$ and (d) what other type of spectrum might help to confirm ethanol as the compound giving this mass spectrum.

2   For methanol suggest and explain how many peaks would be in the low resolution NMR spectrum and the ratio of their areas.

Industrial chemistry: oil refining and petrochemicals

Examiner's secrets

You don't need to to learn distillation and cracking flow diagrams by heart but you should know and understand the principles and differences of the processes.

Watch out!

Saturated alkanes crack to a mixture of alkanes, hydrogen and alkenes (or other unsaturated hydrocarbons).

The jargon

A *fluidized bed* is a fine powder behaving like a liquid when gas passes through.

We depend on petroleum (*rock oil*) for energy and a vast range of organic chemicals. Most of this non-renewable resource is refined and used as fossil fuels. The rest is processed into the raw materials for the chemical industry.

Oil refining ●●●

→ *First stage*: distillation to separate crude oil into fractions:

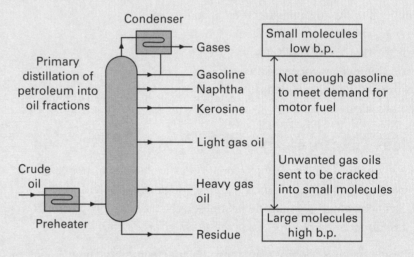

→ *Second stage*: thermal, steam or catalytic cracking to break large molecules into small ones, e.g. $C_{14}H_{30} \rightarrow C_{10}H_{22} + 2C_2H_4$.
→ Cracking of petroleum fractions is a major source of ethene, propene and hydrogen for the chemical industry.
→ In catalytic crackers hot hydrocarbon vapour flows through fluidized beds of *heterogeneous catalysts*

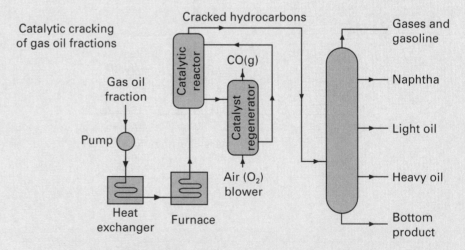

→ *Third stage*: isomerization and reforming.
→ Isomerization is the rearrangement of a molecule into an isomer: e.g. $CH_3CH_2CH_2CH_3 \rightarrow (CH_3)_3CH$ using Pt on Al_2O_3 catalyst.
→ Reforming is the conversion of a molecule into a different type: e.g. $CH_3(CH_2)_5CH_3 \rightarrow C_6H_5CH_3 + 4H_2$.

Chemicals from petroleum ●●●

The following is a selection of petrochemicals and related products you may well encounter in your A-level course.

Ethene

chloroethene: $CH_2=CH_2Cl$ (to make PVC); epoxyethane: $CH_2\overset{O}{-}CH_2$;
ploy(ethene): $-(CH_2-CH_2)_n-$; ethyl benzene: $C_6H_5CH_2CH_3$;
ethane-1,2-diol: CH_2OHCH_2OH (antifreeze and to make polyesters).

Propene

propan-2-ol: $CH_3CH(OH)CH_3$; poly(propene): $-(CH_2-CH(CH_3))_n-$;
acrylonitrile: $CH_2=CH_2CN$ (to make acrylic plastics); cumene:
$C_6H_5CH(CH_3)_2$ (to make phenol and propanone) – see below.

Butadiene

$CH_2=CH-CH=CH_2$ (to make synthetic rubber and other products).

Alicyclics

cyclohexane C_6H_{12} and cyclohexanol $C_6H_{11}OH$ (used to make
1,6-hexanedioic acid and caprolactam for nylon manufacture).

Aromatics

benzene: $C_6H_6 \rightarrow C_6H_{12}$ cyclohexane (to make nylon-6)

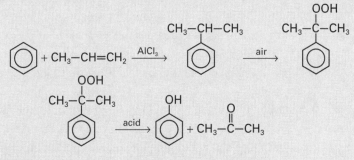

methylbenzene (toluene): $C_6H_5CH_3$ (to make TNT)

Azo-dyes
methyl orange

Polymers
poly(phenylethene)
polystyrene

'Terylene'

Carbon black
fine particles of the element (in printing inks and tyre manufacture)

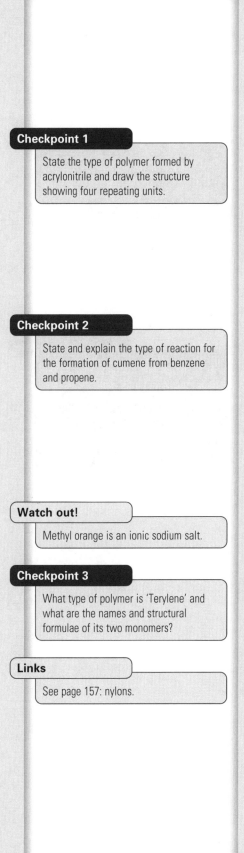

Checkpoint 1

State the type of polymer formed by acrylonitrile and draw the structure showing four repeating units.

Checkpoint 2

State and explain the type of reaction for the formation of cumene from benzene and propene.

Watch out!

Methyl orange is an ionic sodium salt.

Checkpoint 3

What type of polymer is 'Terylene' and what are the names and structural formulae of its two monomers?

Links

See page 157: nylons.

Exam question (5 min)   answer: page 172

(a) In the catalytic cracking of oil suggest (i) why the catalyst needs to be regenerated, and (ii) what might occur in the catalyst regenerator.

(b) What type of molecule (class of compound) is represented by
(i) $CH_3(CH_2)_5CH_3$ and (ii) $C_6H_5CH_3$?

Comprehension question (30 min)

answer: page 172

Read the passage below and then answer the questions (a) to (h) which are based on it.

Examiner's secrets

Don't be put off by complicated structural formulae. Look for functional groups and put a circle around each one with a pencil or pen.

Examiner's secrets

(1) *Skim quickly* through the passage to get the gist of it (2) *glance quickly* at the questions and then (3) read through the passage carefully. If you are told to summarize the passage, underline key points as you read through carefully.

Pain Killing Drugs

Pain killing or analgesic drugs are in common use both for minor aches and pains and for control of more intense pain in serious medical conditions. The structures of five well-known pain killing drugs are shown below.

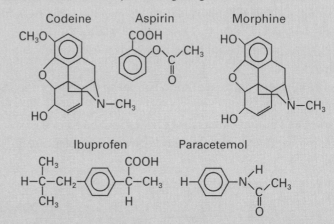

Aspirin and its soluble sodium salt are regarded as the safest of all these drugs, having been in widespread use for so long. Aspirin is made from salicylic acid (2-hydroxybenzenecarboxylic acid), and was first synthesized in 1899. Salicylates, which occur naturally, e.g. in the willow tree, have been used in herbal medicine for a very long time. The salicylates were identified as painkillers in 1876. Aspirin is used for mild pain relief but also reduces inflammation so is used for relief of the symptoms of rheumatoid arthritis in some patients. It is known that, in some cases, aspirin may cause harmful bleeding in the stomach.

Codeine and morphine belong to the opium alkaloids. *Alkaloids* are basic nitrogen-containing naturally occurring compounds, which have pronounced physiological effects on the body. Morphine is the main constituent of opium, a substance derived from the opium poppy. Codeine is a more effective painkiller than aspirin and does not have the addictive and other side effects of morphine. Morphine is used to produce heroin (or *diamorphine*, as it is known in medicine). In heroin the two hydroxy groups of the morphine molecule have been ethanoylated. Heroin is highly addictive and is used mainly for controlling pain in terminally ill patients. Its illegal use and detrimental addictive properties are well documented.

Paracetamol is a readily avaliable painkiller, which, like aspirin, requires no prescription. It is used for mild pain such as headache, toothache and muscle pain. Like all these drugs it is important not to exceed the recommended dosage. Too high a dose of paracetamol can cause harmful physiological effects in the body.

Ibuprofen is the most recent of these drugs. It may be said to be a 'designer drug', having been specifically developed by the Boots Company to alleviate the symptoms of rheumatoid arthritis. Like all drugs that are introduced, it had to undergo extensive testing before approval and before it was available on prescription in 1969. It is now readily available both as non-proprietary ibuprofen or under a variety of proprietary names such as Nurofen.

(a) State which *one* of the five drugs is considered the safest. Give a reason for your answer.

....

....

(b) (i) Draw the structure of 2-hydroxybenzenecarboxylic acid.

(ii) State the number of moles of sodium hydroxide which reacts with one mole of aspirin to form the sodium salt.

....

(c) State which of the drugs shown are classified as *alkaloids*.

....

(d) Complete the molecular skeleton given below to show the structure of a molecule of *diamorphine*.

(e) (i) State the type of isomerism shown by ibuprofen.

....

(ii) Show on the structural formula below the structural feature giving rise to this isomerism.

(f) State which of the five drugs may be classified as a 'designer drug' and give your reasons.

....

....

....

(g) By considering the structure of morphine, explain why

(i) in acidic solution the nitrogen atom becomes protonated

....

....

(ii) in alkaline solution the molecule only loses one proton.

....

....

(h) By referring to the drugs listed give

(i) *two* advantages to society by the provision of such chemicals

....

....

....

(ii) *one* precaution that must be taken when using thee analgesic drugs.

....

....

Structured exam question

answer: page 172

(a) The molecular formulae of six compounds, A, B, C, D, E and F, are shown in the reaction scheme below. Compound A is a colourless liquid that gives a colourless gas with aqueous sodium carbonate. Compound A can be converted directly to E by refluxing with ethanol and concentrated sulphuric acid.

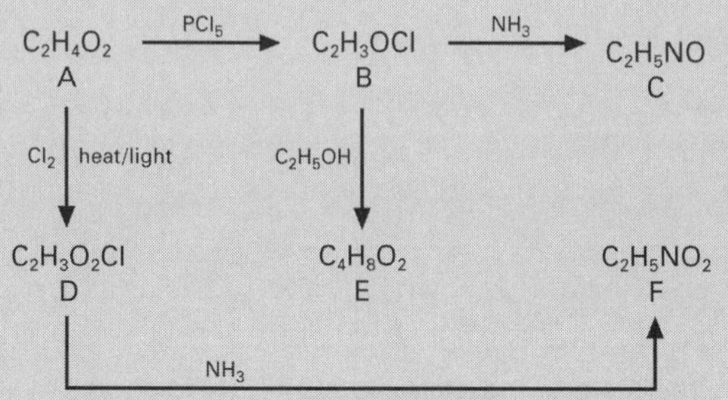

$$C_2H_4O_2 \xrightarrow{PCl_5} C_2H_3OCl \xrightarrow{NH_3} C_2H_5NO$$
$$A \qquad\qquad B \qquad\qquad C$$

Cl_2 | heat/light $\qquad$ C_2H_5OH

$$C_2H_3O_2Cl \qquad C_4H_8O_2 \qquad C_2H_5NO_2$$
$$D \qquad\qquad E \qquad\qquad F$$

NH_3

(i) Draw structural formulae for compounds A to F.

A B C

D E F

Write a balanced equation for the reaction of compound A with

(ii) aqueous sodium carbonate:

....

(iii) ethanol:

....

(iv) Draw a possible structure for a compound formed when F reacts with 2-aminopropanoic acid, $CH_2CH(NH_2)COOH$.

(b) (i) Describe and explain the S_N2 mechanism for the hydrolysis of 1-bromobutane by aqueous sodium hydroxide.

(ii) Write the rate equation for this S_N2 hydrolysis reaction.

....

Links

See pages 140–1: functional groups.

Examiner's secrets

Play safe and show *all* the bonds.

Watch out!

This is one of the very small number of mechanisms required at A-level. Know your mechanisms well!

Answers
Organic chemistry

How to name compounds

Checkpoints

1 (a) 2-methylbutane   (b) 2,3-dimethylheptane

2 hex-2-ene

Three isomers

hex-1-ene

hex-2-ene

hex-3-ene

3 1,3,5-trimethylbenzene

There are four

1,2,3-trichlorobenzene

1,2,4-trichlorobenzene

1,3,4-trichlorobenzene

1,3,5-trichlorobenzene

Exam question

(a) They are saturated hydrocarbons having the same molecular formula but different structures and properties.

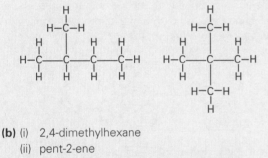

(b) (i)   2,4-dimethylhexane

(ii)   pent-2-ene

(iii)   4-bromo-2,6-dichloromethylbenzene

(c) (i) $CH_3CH_2CH_2CH_2CH_2CH_2CH_2CH_3$

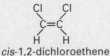

(d)

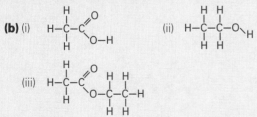

1,1-dichloroethene   *trans*-1,2-dichloroethene

cis-1,2-dichloroethene

Classes of compounds and functional groups

Exam question

(a) nitrile, alkene, phenyl, chloro.

(b) (i)   (ii)   (iii)

(c) $CH_3COOH + C_2H_5OH \rightleftharpoons CH_3COOC_2H_5 + H_2O$

Types of reactions and reagents

Checkpoints

1 (i)   Electrophile

(ii)   Free radical

(iii)   Nucleophile

(iv)   Nucleophile

2

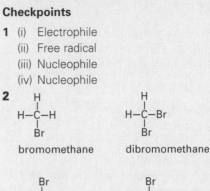

bromomethane   dibromomethane

tribromomethane   tetrabromomethane

3 (a) (i)   An electrophile

(ii)   A nucleophile

(b) 1,2-dibromoethane.

The mechanism is electrophilic addition:

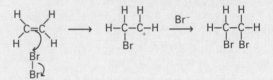

Exam question

(a) (i) $Cl_2 + h\nu \rightarrow 2Cl\cdot$ initiation

(ii) $C_2H_6 + Cl\cdot \rightarrow C_2H_5\cdot + HCl$

$C_2H_5\cdot + Cl_2 \rightarrow C_2H_5Cl + Cl\cdot$ propagation

(iii) $C_2H_5\cdot + C_2H_5\cdot \rightarrow C_4H_{10}$ termination

(iv) $C_2H_6 + 2Cl_2 \rightarrow C_2H_4Cl_2 + 2HCl$

(b) The Cl–Cl bond energy is greater than the Br–Br bond energy. Therefore the photon energy, $h\nu$, to break the Cl–Cl bond must be higher than that to break the Br–Br bond. To get a higher photon energy, the frequency (ν) must be greater (h is Planck's constant).

(c) CFCs such as CCl_2F_2 undergo homolysis (homolytic fission) and form chlorine free radicals. The carbon–fluorine bonds do not break because of the very high bond energy of C–F:

$$CCl_2F_2 + h\nu \rightarrow \cdot CClF_2 + Cl\cdot$$

The chlorine free radicals react with ozone molecules

$$O_3 + Cl\cdot \rightarrow O_2 + ClO\cdot$$

The new radical, $ClO\cdot$, enters into a variety of other radical reactions but the net effect is that ozone molecules are destroyed.

The use of CFCs has caused the appearance of holes in the ozone layer in the upper atmosphere. CFCs reach the stratosphere without degradation because they are very stable compounds.

The ozone layer is important because it absorbs harmful UV radiation. Scientists think that depletion of the ozone layer is causing an increase in diseases such as skin cancer by allowing more harmful UV radiation to reach the earth's surface. The EU has now banned the manufacture of CFCs.

Hydrocarbons: alkanes and alkenes

Checkpoint

Each carbon atom in the ethane molecule is surrounded by four bonding pairs of electrons. The arrangement of the bonds around each carbon atom must be tetrahedral.

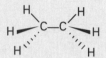

Note that, because there is free rotation about the carbon–carbon bond, the most stable shape is when the end-on view of the molecule is as shown.

This is called the staggered conformation.

Exam question

(a) $NaOH + C_4H_9Br \rightarrow C_4H_9OH + NaBr$ is a substitution reaction in which Br is replaced by OH.
The species which attacks the halogenoalkane (C_4H_9Br) is the hydroxide ion, OH^-. The hydroxide ion is a lone pair donor (a Lewis base) or nucleophile. Thus the above reaction is said to be a nucleophilic substitution reaction.

In an addition reaction, a molecule adds on atoms to become a different molecule. The reaction $CH_2=CH_2 + Br_2 \rightarrow CH_2BrCH_2Br$ is an addition reaction. The mechanism involves the heterolysis (heterolytic fission) of the Br–Br bond. This results in the addition of Br^+ to form $CH_2BrCH_2^+$, a carbocation (carbonium ion). Since Br^+ is an electrophile (a lone pair acceptor or Lewis acid) the reaction is called electrophilic addition.

(b) In the absence of UV light.

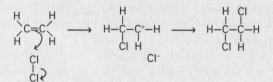

(c) CH_2BrCH_2Br 1,2-dibromoethane
CH_2BrCH_2OH 2-bromoethanol
CH_2ClCH_2Br 1-bromo,2-chloro-ethane

Hydrocarbons: alkenes and arenes

Checkpoints

1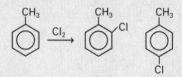

2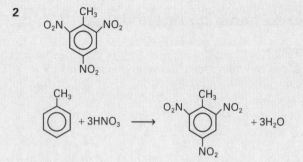

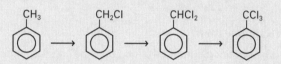

Exam question

(a) (i) When chlorine is passed into hot methylbenzene, then the side-chain is substituted in a free radical reaction.

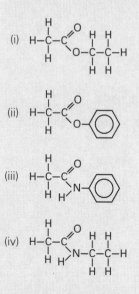

(ii) In the absence of heat and light and in the presence of a halogen carrier, a mixture of 2-chloromethylbenzene and 4-chloromethylbenzene is formed.

(b) This reaction is a free radical reaction.
$$Cl_2 + h\nu \rightarrow 2Cl\cdot \text{ initiation}$$
$$C_6H_5CH_3 + Cl\cdot \rightarrow C_6H_5CH_2\cdot + HCl$$
$$C_6H_5CH_2\cdot + Cl_2 \rightarrow C_6H_5CH_2Cl + Cl\cdot \text{ propagation}$$
Further reaction will occur until all three hydrogen atoms of the side chain have been substituted.

Compounds containing halogens

Checkpoint

(i)
(ii)
(iii)
(iv)

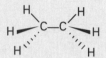

Exam question

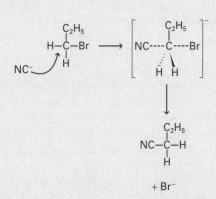

$+ Br^-$

Compounds containing oxygen: alcohols, phenols, aldehydes and ketones

Checkpoints

1

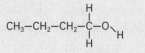

butan-1-ol (primary)

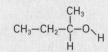

butan-2-ol (secondary)

2-methylpropan-1-ol (primary)

2-methylpropan-2-ol (tertiary)

2 $C_6H_5OH + NaOH \rightarrow C_6H_5O^- + Na^+ + H_2O$

Exam question

(a) Under mild conditions ethanol is oxidized to ethanal.

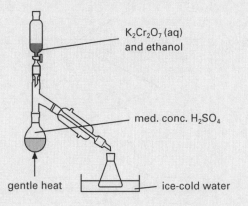

K₂Cr₂O₇ (aq) and ethanol

med. conc. H₂SO₄

gentle heat — ice-cold water

Under harsher conditions, where the reagents are refluxed together, ethanol is oxidized to ethanoic acid.

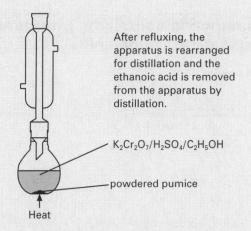

After refluxing, the apparatus is rearranged for distillation and the ethanoic acid is removed from the apparatus by distillation.

$K_2Cr_2O_7/H_2SO_4/C_2H_5OH$

powdered pumice

Heat

(b) (i) Ethene and steam at 70 atm pressure are passed over a catalyst of phosphoric acid on a silica support at 300 °C:
$H_2O(g) + C_2H_4(g) \rightarrow C_2H_5OH(g)$

(ii) When oils or fats are refluxed with aqueous sodium hydroxide, hydrolysis of the ester linkages takes place and results in a mixture of the sodium salts of carboxylic (fatty) acids and glycerol (propan-1,2,3-triol).

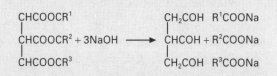

Compounds containing oxygen: aldehydes, ketones and carboxylic acids

Checkpoints

1

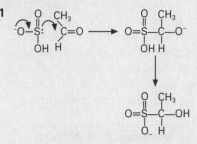

2 $C_2H_5COOH + CH_3OH \rightleftharpoons C_2H_5COOCH_3 + H_2O$

Exam question

(a)

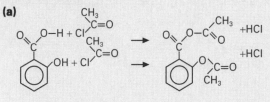

(b) Reflux a mixture of 2-hydroxbenzoic acid with methanol and a small amount of concentrated sulphuric acid to catalyze the reaction and remove water to shift the equilibrium to the right to improve the yield.

Compounds containing nitrogen: amines and amino acids

Checkpoints

1 (i)

2-aminobutane – primary

(ii)

NH₂ ... CH₃ (benzene ring)

2-methylphenylamine – primary

(iii)

CH_3
$>N-C_2H_5$
CH_3

ethyldimethylamine – tertiary

(iv)

$CH_3 \quad CH_3$
N (on benzene ring)

N:N-dimethylphenylamine – tertiary

2 $CO(NH_2)_2 + 2HNO_2 \rightarrow CO_2 + 3H_2O + 2N_2$

Note: You might expect the product to be $CO(OH)_2$. This is carbonic acid, H_2CO_3, which decomposes to carbon dioxide and water.

3

Br, Br, Br substituted on NH₂-benzene ring

Exam question

(a) Phenylamine is only very slightly soluble in water but the salt, $C_6H_5NH_3Cl$, is readily soluble.

(b) The alkali reacts with the phenolic group to give the soluble sodium salt whereas 2-naphthol is insoluble.

(c) Aromatic diazonium ions are stabilized by electrons of the multiple bond between the nitrogen atoms participating in the delocalization of the π-electrons of the ring.

Compounds containing nitrogen: proteins and polyamides

Checkpoints

1 $(CH_3CO)_2O + 2NH_3 \rightarrow CH_3COONH_4 + CH_3CONH_2$

2 $n\text{ClOC}(CH_2)_8\text{COCl} + n\text{H}_2\text{N}(CH_2)_6\text{NH}_2$

$\downarrow$

$-[\text{NHOC}(CH_2)_8\text{CONH}(CH_2)_6]_n- + 2n\text{ HCl}$

Exam question

(a) Tin, concentrated hydrochloric acid and nitrobenzene are refluxed together in an apparatus contained within a fume cupboard. When the reaction is complete, the mixture is allowed to cool. At this stage, the phenylamine is in the

form of phenylammonium hexachlorostannate(IV). Sodium hydroxide is added to make the mixture alkaline and release the phenylamine, which is removed by steam distillation.
Overall
$3Sn + 2C_6H_5NO_2 + 14HCl \rightarrow$
$(C_6H_5NH_3)_2SnCl_6 + 2SnCl_4 + 4H_2O$
The addition of sodium hydroxide releases the phenylamine:
$C_6H_5NH_3^+ + OH^- \rightarrow C_6H_5NH_2 + H_2O$

(b) At temperatures greater than 10 °C, phenol is formed:
$C_6H_5NH_2 + HNO_2 \rightarrow C_6H_5OH + N_2 + H_2O$
Between 0 °C and 10 °C, benzenediazonium chloride is formed:
$C_6H_5NH_2 + 2HCl + NaNO_2 \rightarrow C_6H_5N_2Cl + NaCl + H_2O$

(c)

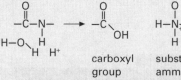

(d) (i) Ammonia is readily evolved:
$RCOONH_4 + NaOH \rightarrow RCOONa + NH_3 + H_2O$

   (ii) Primary amines do not react with aqueous sodium hydroxide.

   (iii) Ammonia is given off on heating:
   $RCONH_2 + NaOH \rightarrow RCOONa + NH_3$

(e) Since nylon is a polyamide, on refluxing with hydrochloric acid the peptide links would be hydrolysed.

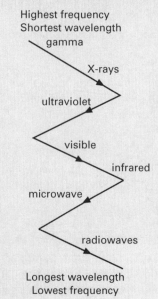

carboxyl group   substituted ammonium group

Instrumental techniques: UV, visible and IR spectroscopy

Checkpoints

1

Highest frequency
Shortest wavelength
gamma
X-rays
ultraviolet
visible
infrared
microwave
radiowaves
Longest wavelength
Lowest frequency

2 (a) The blue end. Bluish light is absorbed so that the pinkish light is transmitted.

(b) You to would be expected to know that chlorophyll is green and therefore to deduce that red and blue light must be absorbed giving two peaks in the red and blue regions. Here is an absorption spectrum of chlorophyll:

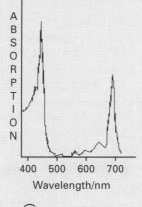

3

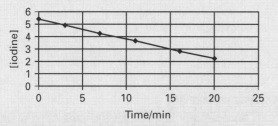

Exam questions

1 The 2nd, 3rd and 5th readings are at 4-min intervals and correspond to the same change in concentration of 0.64×10^{-4} mol dm^{-3}. This suggests concentration varies linearly with time.

The graph of [iodine] against time is a straight line, so the reaction is zero order with respect to iodine.

2 (a) A sample of the contaminated PABA would be scanned in a UV/visible spectrometer and the instrument's computer would compare the result with its database spectrum for pure PABA. A trace contaminant would probably generate unexpected peaks in the sample's spectrum. An IR spectral analysis would be run in the same way.

(b) Samples of exhaust gases (a complex mixture of gases including carbon monoxide, carbon dioxide, nitrogen and oxides of nitrogen together with unburned hydrocarbons and hydrocarbon degradation products) would be withdrawn from within the exhaust system and analysed
1 by separating the components using gas/liquid chromatography and then
2 analysing the separate components by infrared spectroscopy.

Instrumental techniques: NMR and mass spectrometry

Checkpoint

The instrument is at low pressure
- to prevent the formation of spurious ions
- so that the ions have an unhindered passage through the apparatus.

Exam questions

1 (a) 31 could be CH_2OH^+
29 could be $C_2H_5^+$
(b) 17 OH^+
15 CH_3^+
(c) A parent molecule ion containing a ^{13}C isotope
(d) Infrared

2 Two peaks in the ratio 3 : 1.
There are two types of proton. The ^1H of the $-OH$ group and the three ^1H of the methyl group. The methyl protons provide a peak which is three times the area of that due to the hydroxyl proton.

Industrial chemistry: oil refining and petrochemicals

Checkpoints

1 Polyacrilonitrile is an addition polymer:

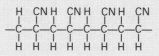

2 When benzene and propene combine, an electrophilic substitution reaction takes place. This is often called a Friedel–Crafts reaction.

Charles Friedel (French) 1832–1899
James C. Crafts (American) 1839–1917

3 A condensation polymer.
Monomers are
1,4-benzenedicarboxylic acid
ethane-1,2-diol

Exam question

(a) (i) The efficiency of the catalyst falls over a period of time and so it is recycled through a regeneration process to restore its efficiency and to contribute to the economy of the process.

(ii) Hydrocarbon impurities are oxidized.

(b) (i) an alkane (saturated)

(ii) an arene (unsaturated)

Comprehension question

(a) Aspirin has been in use for a long time.

(b) (i)

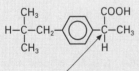

(ii) One

(c) Codeine and morphine

(d)

(e) (i) Optical isomerism

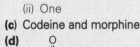

(ii) Asymmetric carbon atom clearly indicated.

(f) Ibuprofen was specially produced to treat rheumatoid arthritis.

(g) (i) The nitrogen atom has a lone pair of electrons to combine with H^+.

(ii) Only the phenolic —OH group is acidic; the other —OH group is alcoholic/is not phenolic group.

(h) (i) This is a fairly open ended question so we would give you marks for any reasonable comments referring to

advantages to society: e.g. control of severe pain in serious medical conditions, to make patient more comfortable, to enable minor discomfort not to affect the quality of daily life, to prevent loss of working day by alleviating minor aches and pains. Use of non-prescription drugs for self-medication saving GP's time, etc.

(ii) Not to exceed the stated dose could be linked to addictive nature of some drugs.

Structured exam question

(a) (i)

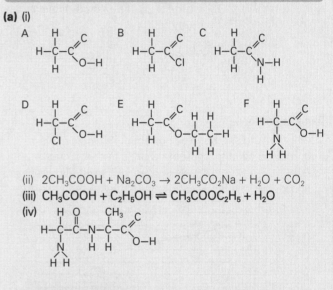

(ii) $2CH_3COOH + Na_2CO_3 \rightarrow 2CH_3CO_2Na + H_2O + CO_2$

(iii) $CH_3COOH + C_2H_5OH \rightleftharpoons CH_3COOC_2H_5 + H_2O$

(iv)

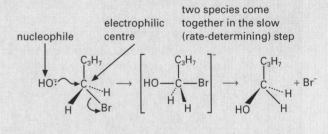

(b) (i) Substitution nucleophilic bimolecular:

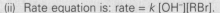

(ii) Rate equation is: rate = k [OH⁻][RBr].

Resources section

This section is intended to help you develop your study skills for exam success. You will benefit if you try to develop skills from the beginning of your course. Modern A-level exams are not just tests of your recall of textbooks and your notes. Examiners who set and mark the papers are guided by assessment objectives that include skills as well as knowledge. You will be given advice on revising and answering questions. Remember to practise the skills.

Exam board specifications

In order to organize your notes and revision you will need a copy of your exam board's syllabus specification. You can obtain a copy by writing to the board or by downloading the syllabus from the board's website.

AQA (*Assessment and Qualifications Alliance*)
Publications Department, Stag Hill House, Guildford, Surrey GU2 5XJ – www.aqa.org.uk

CCEA (*Northern Ireland Council for Curriculum, Examinations and Assessment*)
Clarendon Dock, 29 Clarendon Road, Belfast BT1 3BG – www.ccea.org.uk

EDEXCEL
Stewart House, 32 Russell Square, London WC1B 5DN – www.edexcel.org.uk

OCR (*Oxford, Cambridge and Royal Society of Arts*)
1 Hills Road, Cambridge CB2 1GG – www.ocr.org.uk

WJEC (*Welsh Joint Education Committee*)
245 Western Avenue, Cardiff CF5 2YX – www.wjec.co.uk

Topic checklist

○ AS ● A2	AQA	CCEA	EDEXCEL	OCR	WJEC
Study skills	○●	○●	○●	○●	○●
Synoptic assessment: syllabuses and exams	○●	○●	○●	○●	○●
Data handling 1: calculations and problem solving	○●	○●	○●	○●	○●
Data handling 2	○●	○●	○●	○●	○●

Study skills

Check the net

You can find a wealth of information on the internet. Here are a few places to start looking:
www.chemweb.com
www.chemsoc.org
www.acdlabs.com/download

The jargon

The *world wide web* (www) is a dynamic system of interconnected computers situated around the world (the *internet*). It is not static and websites are changing all the time.

Check the net

Visit the sites of colleges, universities and institutes of higher education before attending an open day. Don't restrict yourself to UK sites.

Study techniques

How good are your study skills?
Answer these questions

1 Do you enjoy studying chemistry?   yes ☐ no ☐
2 Do you have good study habits?   yes ☐ no ☐
3 Do you follow a programme of spaced revision?   yes ☐ no ☐
4 Do you regularly tackle chemistry questions?   yes ☐ no ☐
5 Do you take practice papers under exam conditions?   yes ☐ no ☐

Smart study

We are usually better at subjects we like than at subjects we dislike. The more we like a subject the easier we find it and the more time we spend studying it. A habit is something that doesn't need great effort or will power. Good students never seem to have to think about studying because they habitually use good study techniques. Top students

→ do private study in the same place at the same times each day
→ warm up at the start of each study session with a simple task
→ work up gradually from easier to more difficult parts of a topic
→ stop each study session while they are still enjoying success
→ clear their table and leave a simple task to start their next session
→ never leave revision to the last minute

Various investigations and research have shown that top students always revise new material as soon as possible and they use frequent and short periods of revision that are carefully spaced. Tony Buzan, an expert on study techniques and author of the book *Make the Most of Your Mind*, recommends a pattern of spaced revision to follow a one-hour study session:

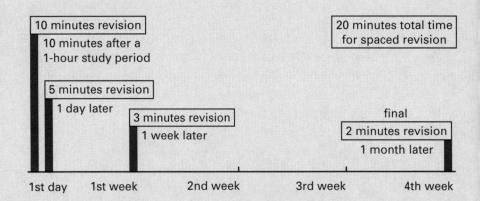

Really good students know the benefit of preparing for examinations by regularly tackling questions and practice papers and then correcting their answers very carefully. In this way you get to know the language the examiners use and have a much clearer idea of what the examiners want for a good answer.

What are examiners looking for?

Questions are set for you to show your knowledge and understanding of fundamental facts, patterns, principles and theories of chemistry and to demonstrate that you can

→ write and balance equations
→ do calculations (pH, ΔH, K_c, E, etc.)
→ predict the feasibility of reactions
→ deduce organic structures

In the synoptic assessment you can draw upon your knowledge and understanding of all the individual modules to reveal your grasp of AS and A2 chemistry as a whole. When you answer the questions you will inevitably display your *quality of written communication*.

To help you understand the questions, here is a list of the more common words and phrases examiners use.

Concise answers with the bare minimum of detail
Classify each of the following oxides as acidic, basic or amphoteric.
Define the term molar first ionization energy.
Give the oxidation number of uranium in the compound.
Indicate the conditions needed to increase the equilibrium yield.
Name the mechanism for ammonia reacting with bromoethane.
State Hess's law.
What is meant by a buffer solution?
Write a balanced equation for the complete combustion of ethanol.

Concise answers with essential but rather more detail
Calculate the activation energy from the data provided.
Comment on the difference in physical properties of CO_2 and SiO_2.
Deduce the structure of the compound from the information provided.
Draw a labelled Born–Haber cycle for the formation of calcium oxide.
Identify the compounds X, Y and Z in the following observations.
Outline a laboratory method of measuring a named enthalpy change.
Show how you would detect the presence of sodium in a compound.
Sketch the unit cell of a body-centred cubic structure.

Longer answers reasoned with facts and principles
Explain why ammonia is basic and forms complexes with cations.
Explain what is meant by fractional distillation.
State and explain the effect of temperature upon reaction rates.
Suggest how to distinguish 1-bromobutane from 2-bromobutane.

In the synoptic elements of your examination you could find all these types of question. The synoptic papers in year A2 will link together the modules you have studied to assess your overall grasp of chemistry.

Synoptic assessment: syllabuses and exams

The risk with syllabuses and exams based on modules is that you might learn each module and pass each modular exam without connecting the parts and seeing the subject as a whole. The synoptic assessment was introduced for you to show examiners your overall grasp of a subject.

The jargon

Greek *syn* – together, *opsis* – seen. *Synoptic chart* is a weather map showing details of temperature, pressure, amount of cloud, etc., so they can be 'seen together'.

Check the net

Visit the Qualifications and Curriculum Authority website at www.qca.org.uk

Exam board sites:
www.ccea.org.uk
www.edexcel.org.uk
www.neab.ac.uk
www.ocr.org.uk
www.wjec.co.uk

The jargon

Subject criteria are the instructions to the exam boards and examiners responsible for the papers you take.

A-level syllabuses and synoptic assessment

Subject criteria

The Qualifications and Curriculum Authority (QCA) has published (on the web) subject criteria for A-level Chemistry specifying that synoptic assessment should

→ require candidates to make connections between different areas of chemistry, for example, by applying knowledge and understanding of principles and concepts of chemistry in planning experimental work and in analysis and evaluation of data.
→ include opportunities for candidates to use, in contexts which may be new to them, skills and ideas that permeate chemistry, for example, writing chemical equations, quantitative work, relating empirical data to knowledge and understanding.

You will find subject criteria included in the assessment objectives of your exam syllabus. Visit your exam board website.

How will synoptic assessment work?

Exam boards may have slightly different assessment objectives and ways of meeting them but they all comply with the following QCA requirements:

1 synoptic assessment will apply to A2 Chemistry
2 synoptic assessment will be at least 20% of the examination

You would normally expect to take written synoptic question papers at the end of the second year of your A-level Chemistry course.

How will synoptic assessment affect you?

According to QCA the emphasis of synoptic assessment is on understanding and application of principles. This means you must be able to

1 bring together your knowledge and understanding of principles and concepts from different areas of chemistry (including your laboratory practical work) and apply them in a particular context.
2 express your ideas clearly and logically using the appropriate specialist language (the jargon) of chemistry.

What should you do?

→ Don't be surprised to find unfamiliar topics in a question
→ Try to apply what you know and understand to new situations
→ Prepare yourself by practising synoptic chemistry questions
→ Get help from your teachers and your fellow students

"A journey of a thousand miles must begin with a single step."

Chinese proverb

A good exam technique can make the difference between one grade and another. For the best grade you must be sure to keep on top of your work throughout your course.

The examination

Preparing for and taking exams

→ Check the regulations to see what you can have in the exam room.

For some papers you may be given a data sheet as well as the periodic table. You may have to use the *Nuffield Advanced Science Book of Data*. This is particularly helpful when you are doing calculations in physical chemistry.

→ Do not use an 'erasable pen' and do not use white correcting fluid.

If you think you have made a mistake, cross it out neatly with one ruled line. Examiners look at your 'mistakes' and may sometimes be allowed to award you marks for what you have crossed out.

→ Tackle at least one set of past or specimen papers.

Use the number of marks and the time allowed for each paper to calculate the *mark rate*: it is often about 1 mark per minute.

→ Take off your wristwatch, put it in front of you where you can see it clearly and get into the habit of checking to keep to time.

If you are running out of time towards the end of an exam, abandon sentences and write your answers in note form even if you run the risk of losing marks for quality of written communication.

The results

We all know how much your results could affect your life. Most students get the result they deserve. If you find that your result is unexpected, you can appeal to the exam board and ask for

1 a clerical check (have marks been added up and entered into the computer correctly)
2 a re-mark (usually by the Chief Examiner or a senior examiner)
3 a re-mark with a report

In a pilot scheme in 1999 some students were given back their marked scripts and their schools and colleges received the marking schemes. If this practice is extended to all boards and subjects, appeals upon results may become unnecessary.

A final word

Self-motivation is the key and is as important as ability. This book, your teachers and fellow students can help but your success depends upon you. Set yourself achievable goals. Develop good study habits. Practise your exam techniques. Work hard. Play hard. And above all, enjoy your chemistry.

Good luck!

Watch out!

Calculators must be silent, cordless and unable to store and display text or graphics. The memory of a programmable calculator must be cleared. Calculator instruction manuals are not allowed into the exam room.

Examiner's secrets

Do not spend too long on one question and not enough time on another. Time lost can rarely be recovered. Misjudging the time is one of the common mistakes you must avoid. *Never* use white correcting fluid because your script would not be valid for an appeal.

Watch out!

Appeals cost money and don't always produce the result you want!

Data handling 1: calculations and

At A-level you will be given data to interpret and convert from one form into another. The data will usually be in the form of tables, diagrams and graphs but sometimes it may just be information described in a few sentences.

Physical constants

Physical constants have a name, a symbol, a value and units:

Avogadro constant	L	6.02×10^{23}	mol^{-1}	per mole
elementary charge	e	1.602×10^{-19}	C	coulomb
gas constant	R	8.314	$J\ K^{-1}\ mol^{-1}$	joule per Kelvin per mole
Faraday constant	F	96 500	$C\ mol^{-1}$	coulomb per mole
Molar gas volume	V_m	22 400	cm^3	cubic centimetre

> **Exam question 1** answer: page 184
>
> A direct current of 0.386 A passed through dilute aqueous sulphuric acid for 10.0 s produced a tiny bubble of hydrogen at an inert cathode. Calculate (a) the amount of charge passed in coulombs, (b) the volume of the hydrogen bubble at 273 K and 1 atm, and (c) the number of H_2 molecules in the bubble. [1 A s = 1 C]

Infrared spectra

→ An infrared correlation table lists absorption frequency range, class of compound and group(s) causing the absorption.

When bond stretching or bending cause a molecule to absorb energy, less infrared radiation passes through the sample and a trough appears in the spectrum.

→ Infrared spectra display percentage transmission against frequency.

Wavenumber/cm^{-1}	Class of compound	Bond
3 750–3 200	alcohols and phenols	O–H and N–H
3 095–3 010	alkene and arene	C–H
2 970–2 850	alkane	C–H
2 260–2 100	alkynes and nitriles	C≡C and C≡N
1 740–1 720	aldehyde	C=O
1 700–1 680	ketone	C=O
1 680–1 620	alkenes and arenes	C=C
1 480–1 360	alkenes and arenes	C–H
1 200–1 050	alcohols	C–O
1 200–800	alkanes	C–C
800–600	chloroalkane	C–Cl
600–500	bromoalkane	C–Br

problem solving

Three straight-chain organic compounds, A, B and C, have the infrared spectra shown below. One compound has the molecular formula $C_8H_{18}O$ and the other two are isomeric with the molecular formula $C_7H_{14}O$.

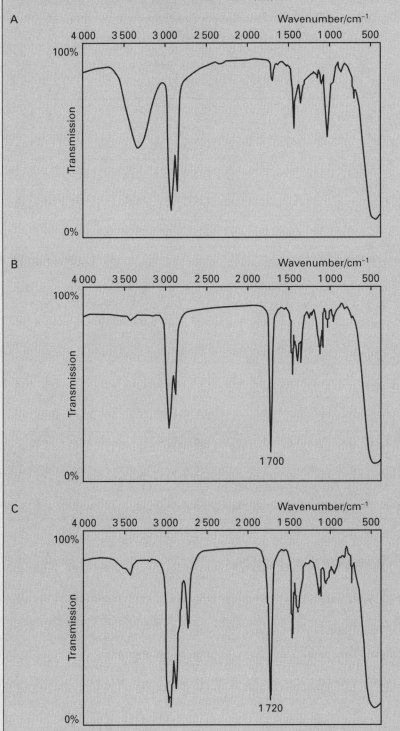

Use the IR correlation data to identify the class of compound and the functional group in A, B and C and write a possible structural formula for each compound.

Data handling 2

Nuclear magnetic resonance

→ A chart of ^1H resonances shows groups containing the hydrogen and the corresponding range of chemical shifts, δ, in ppm.

When resonance occurs (because the radiowave energy from the transmitting oscillator matches the energy difference of the nuclear spin states) the receiver detects a decrease in signal intensity.

→ NMR spectra display absorption against chemical shift (δ).

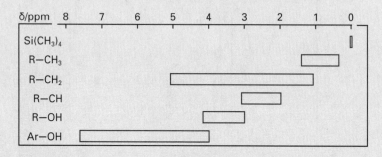

Exam question 3   answer: page 184

Two straight-chain organic compounds, X and Y, gave the NMR spectra shown below. The empirical formula of one is C_2H_6O and that of the other is $C_4H_{10}O$.

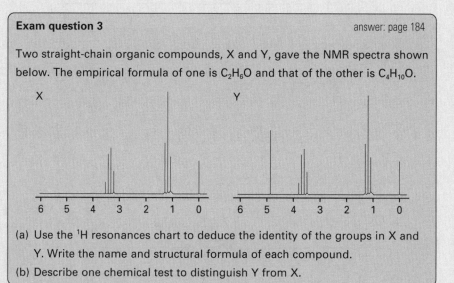

(a) Use the ^1H resonances chart to deduce the identity of the groups in X and Y. Write the name and structural formula of each compound.
(b) Describe one chemical test to distinguish Y from X.

Average bond energies

→ Bond energy (or enthalpy) is the average energy required to break one mole of the specified bond and is always a positive value.
→ Some values and the bond dissociation energies (for diatomic elements) are values applying to specific molecules.

Examiners often give the necessary data with a question but sometimes you have to select what you need from a larger table in your data book.

Bond	Energy/kJ mol⁻¹	Bond	Energy/kJ mol⁻¹
C=O	805	N≡N	945
H—O	464	N=O	470
C—H	413	N—H	388
C—N	286	N—N	158

1,2-dimethylhydrazine, a potential rocket fuel, ignites spontaneously on mixing with dinitrogen tetraoxide, a potential oxidant. The following is a possible equation for the combustion reaction:

$$CH_3-NH-NH-CH_3 + 2N_2O_4 \rightarrow 2CO_2 + 4H_2O + 3N_2$$

Use the table of average bond energies to calculate the energy change for the reaction. Comment briefly on the suitability of a mixture of these two nitrogen compounds as a rocket propellant.

Standard enthalpy changes

Data books list $\Delta H^{\ominus}_{f,298}$ for inorganic and organic compounds and $\Delta H^{\ominus}_{c,298}$ for organic (and some inorganic) compounds. You may be given the necessary values (usually in kJ mol^{-1}) with the question but you may be asked to choose values from your data book.

(a) Use the standard enthalpy changes of combustion (in kJ mol^{-1}) in the table to calculate the standard enthalpy of formation of each of the four cyclic hydrocarbons.

carbon	C	−394	cyclohexadiene	C_6H_8	−3543
hydrogen	H_2	−286	cyclohexene	C_6H_{10}	−3752
benzene	C_6H_6	−3267	cyclohexane	C_6H_{12}	−3920

(b) Use the enthalpies of formation from (a) to calculate the enthalpy change for (i) $C_6H_6 + 3H_2 \rightarrow C_6H_{12}$ and (ii) $C_6H_8 + C_6H_{10} + 3H_2 \rightarrow 2C_6H_{12}$ and comment briefly on the difference between the two values.

Standard redox potentials

→ Data books list the standard e.m.f. of electrochemical cells in which the standard hydrogen electrode forms the left-hand side.
→ Values of $E^{\ominus}/V$ are listed with the right-hand side of the cell and are usually arranged in order from most negative to most positive.

Electrode system	$E^{\ominus}/V$
X $[Cr_2O_7^{2-}(aq) + 14H^+(aq)],[2Cr^{3+}(aq) + 7H_2O(l)] \| Pt$	+1.33
Y $Cl_2(aq), 2Cl^-(aq) \| Pt$	+1.36
Z $[MnO_4^-(aq) + 5H^+(aq)],[Mn^{2+}(aq) + 4H_2O(l)] \| Pt$	+1.51

Use the above data to answer the following questions.

(a) Write the complete cell diagram for the electrochemical cell with a standard e.m.f. of −1.36 V.

(b) What would be the standard e.m.f. of a cell composed of X and Z?

(c) Why do manganate(VII) titrations use sulphuric acid and not hydrochloric acid to acidify the solutions?

(d) What would be the effect, if any, of adding aqueous potassium dichromate(VI) to aqueous manganese(II) sulphate?

Watch out!

In N_2O_4 the bonding is delocalized and the NO bond is not strictly speaking N=O, but for the purposes of this calculation treat it as if it is. See page 99.

A century of chemistry Nobel Prize

The last hundred years have seen great advances in chemistry which have affected our everyday lives.

The contributions of chemistry to society

We live in a chemical age. The advances in chemistry can be summarized in the list of the Nobel Prize Winners for Chemistry through the twentieth century.

Where two or more names appear in the list for a particular year, then the prize was shared.

Checkpoint

Go through the list of winners and mark any whose names appear in the rest of the book.

Chemistry Nobel Prize Winners 1901–1999

Year	Prize winners	Subject
1999	Ahmed Zewail	femtosecond spectroscopy
1998	Walter Kohn, John A. Pople	theoretical chemistry
1997	Paul D. Boyer, John E. Walker, Jens C. Skou	synthesis ATP discovery of an ion-transporting enzyme
1996	Robert F. Curl, Jr., Sir Harold W. Kroto, Richard E. Smalley	discovery of the fullerenes
1995	Paul Krutzen, Mario Molina, F. Sherwood Rowland	ozone depletion in the atmosphere
1994	George A. Olah	carbocation chemistry
1993	Kary B. Mullis, Michael Smith	invention of the polymerase chain reaction contributions to protein studies
1992	Rudolph A. Marcus	electron-transfer reactions
1991	Richard R. Ernst	NMR spectroscopy
1990	Elias James Corey	organic synthesis
1989	Sidney Altman, Thomas R. Cech	catalytic properties of RNA
1988	Johann Deisenhofer, Robert, Huber, Hartmut Michel	three-dimensional structure of a photosynthetic reaction centre
1987	Donald J. Cram, Jean-Marie Lehn, Charles J. Pedersen	structure-specific interactions of high selectivity
1986	Dudley R. Herschbach, Yuan T. Lee, John C. Polanyi	dynamics of chemical elementary processes
1985	Herbert A. Hauptman, Jerome Karle	methods for the determination of crystal structures
1984	Robert B. Merrifield	chemical synthesis on a solid matrix
1983	Henry Taube	electron transfer in metal complexes
1982	Sir Aaron Klug	crystallographic electron microscopy
1981	Kenichi Fukui, Ronald Hoffmann	theories in theoretical chemistry developed independently
1980	Paul Berg	recombinant DNA
	Walter Gilbert, Frederick Sanger	base sequences in nucleic acids
1979	Herbert C. Brown, Georg Wittig	boron and phosphorus chemistry in organic synthesis
1978	Peter D. Mitchell	chemiosmotic theory
1977	Ilya Prigonine	stereochemistry non-equilibrium thermodynamics
1976	William N. Liscombe	the boranes
1975	Sir John Warcup Cornforth,	stereochemistry of enzyme-catalysed reactions
	Vladimir Prelog	stereochemistry of organic molecules
1974	Paul J. Flory	macromolecules
1973	Ernst Otto Fischer, Sir Geoffrey Wilkinson	organometallic sandwich compounds
1972	Christian B. Anfinsen	amino acid sequencing
	Standford Moore, William H. Sein	catalytic activity of the active centre of the ribonuclease molecule
1971	Gerhard Herzberg	structure and geometry of free radicals
1970	Luis F. Leloir	sugar nucleotides
1969	Sir Derek H. R. Barton, Odd Hassel	concept of conformation and its application
1968	Lars Onsager	thermodynamics
1967	Manfred Eigen, Ronald George Norrish, Lord George Porter	fast chemical reactions
1966	Robert S. Mulliken	molecular orbital chemistry
1965	Robert Burns Woodward	organic synthesis
1964	Dorothy C. Hodgkin	X-ray techniques of biological compounds
1963	Karl Ziegler, Giulio Natta	polymer catalysts
1962	Max Ferdinand Perutz, Sir John C. Kendrew	study of globular proteins
1961	Melvin Calvin	CO_2 assimilation in plants

Winners 1901–1999

Year	Prize winners	Subject
1960	Willard Frank Libby	carbon-14 dating
1959	Jaroslav Heyrovsky	polarographic analysis
1958	Frederick Sanger	work on proteins and especially insulin
1957	Lord Alexander Todd	nucleotides and nucleotide co-enzymes
1956	Sir Cyril Hinshelwood, Nikolay Semenov	reaction mechanisms
1955	Vincent du Vigneaud	first synthesis of a polypeptide hormone
1954	Linus C. Pauling	nature of the chemical bond
1953	Hermann Staudinger	macromolecular chemistry
1952	Archer J. P. Martin, Richard L. M. Synge	discovery of partition chromatography
1951	Edwin McMillan, Glenn T. Seaborg	chemistry of the transuranium elements
1950	Otto Paul Diels, Kurt Alder	diene synthesis – the Diels-Alder reaction
1949	William Francis Giauque	low temperature thermodynamics
1948	Arne Wilhelm Kaurin Tiselius	electrophoresis and serum proteins
1947	Sir Robert Robinson	alkaloid chemistry
1946	James Batcheller Sumner, John H. Northrop, Wendell M. Stanley	crystallizing enzymes / pure enzymes and virus proteins
1945	Artturi I. Virtanen	agricultural and nutrition chemistry
1944	Otto Hahn	discovery of nuclear fission
1943	George de Hevesy	isotopes as tracers
1940–42	**No prize awarded**	
1939	Adolf F. J. Butenandt	work on sex hormones
	Leopold Ruzicka	polymethylenes and higher terpenes
1938	Richard Kuhn	carotenoids and vitamins
1937	Sir Walter N. Haworth, Paul Karrer	vitamins A and B2
1936	Petrux J. W. Debye	dipole moments and X-ray diffraction
1935	Frederick Joliot, Irene Joliot-Curie	synthesis of new radioactive elements
1934	Harold C. Urey	discovery of heavy hydrogen
1933	**No prize awarded**	
1932	Irving Langmuir	surface chemistry
1931	Carl Bosch Friedrich Bergius	chemical high pressure methods
1930	Hans Fischer	haemin and chlorophyll
1929	Sir Arthur Harden, Hans von Euler-Chelpin	fermentation processes and enzymes
1928	Adolf Otto Windaus	sterols and vitamins
1927	Heinrich Otto Wieland	bile acids
1926	Theodore Svedberg	disperse systems
1925	Richard A. Zsigmondy	colloid chemistry
1924	**No prize awarded**	
1923	Fritz Pregl	organic microanalysis
1922	Francis W. Aston	mass spectrometry and isotopes
1921	Frederick Soddy	radioactivity
1920	Walther H. Nernst	thermodynamics
1919	**No prize awarded**	
1918	Fritz Haber	synthesis of ammonia
1916–17	**No prize awarded**	
1915	Richard Martin Willstätter	chlorophyll and plant pigments
1914	Theodore William Richards	accurate determinations of atomic weights
1913	Alfred Werner	work in bonding in inorganic chemistry
1912	Victor Grignard,	organometallic chemistry – Grignard reagents
	Paul Sabatier	catalytic hydrogenation
1911	Marie Curie	discovery of radium and polonium
1910	Otto Wallach	chemical industry and alicyclic compounds
1909	Wilhelm Ostwald	catalysis/equilibria/kinetics
1908	Lord Ernest Rutherford	radioactivity
1907	Eduard Buchner	biochemistry/cell-free fermentation
1906	Henri Moissan	fluorine chemistry
1905	Johann von Baeyer	organic chemistry
1904	Sir William Ramsay	discovery of the inert gases
1903	Svante Arrhenius	electrolytic theory
1902	Herman Emil Fischer	sugar and purine syntheses
1901	Jacobus Van't Hoff	chemical dynamics and osmotic pressure

Answers
Data handling

1 (a) The charge passed is the time in seconds multiplied by the current in ampères. Charge = $0.386 \times 10.0 = 3.86$ C.
   (b) No. of Faradays passed = $3.86/96\,500 = 4.00 \times 10^{-5}$ F.
   $$2H^+ + 2e^- \rightarrow H_2$$
   Therefore, 2 Faradays form one mole of hydrogen.
   No. of moles of hydrogen formed = 2.00×10^{-5} mol.
   Thus volume of bubble (at s.t.p.) = $22\,400 \times 2.00 \times 10^{-5}$ cm^3 = 0.448 cm^3.
   (c) No. of molecules of hydrogen in the bubble = $2.00 \times 10^{-5} \times 6.02 \times 10^{23} = 1.20 \times 10^{19}$.

2 Compound A appears to be an alkanol. The absorptions correspond to alkane and alcohol. It is not possible to distinguish between a primary and secondary alcohol. A could be $CH_3(CH_2)_6CH_2OH$.
   Compounds B and C have similar absorptions around 1 700 cm^{-1}. The alkane absorptions suggest that they are aliphatic carbonyl compounds. The carbonyl group absorption for compound B has a slightly larger wavenumber than that in compound C. Therefore compound B is an aldehyde such as $C_6H_{13}CHO$ and compound C is a ketone such as $(C_3H_7)_2CO$.

3 (a) Examination of the two NMR spectra shows that compound Y has a peak corresponding to a single ^{1}H in an —OH group. The remainder of the spectrum of Y is consistent with a quadruplet of lines for —CH$_2$ protons coupling with —CH$_3$ and a triplet for the —CH$_3$ protons. The spectrum is consistent with ethanol, CH_3CH_2OH.
   Compound X is consistent with the presence of a —C_2H_5 group, with the characteristic quadruplet and triplet of lines. Compound X is likely to be ethoxyethane, $(C_2H_5)_2O$.
   (b) There are a number of chemical tests to distinguish the two compounds. Ethoxyethane is classified as an ether and is relatively inert. The —OH group of the alcohol is quite reactive, e.g. ethanol reacts with metallic sodium to give hydrogen, ethoxyethane does not. Ethanol readily reduces acidified potassium dichromate(VI) to give a colour change from orange to green. Ethoxyethane does not. Only ethanol gives the iodoform reaction.

4 Bonds broken (endothermic), ΔH is positive:
   $6(C{-}H) + 2(N{-}H) + 2(C{-}N) + (N{-}N) + 4(N{=}O) + (N{-}N)$
   $= 6\,022$ kJ
   Bonds formed (exothermic), ΔH is negative:
   $4(C{=}O) + 8(H{-}O) + 3(N{\equiv}N) = -9\,382$ kJ
   The enthalpy change for the reaction is $-3\,360$ kJ mol.
   This reaction is extremely exothermic and produces a large volume of gaseous products. This is ideal for jet propulsion since a large volume of hot gases is produced from a relatively small volume of liquid reactants.

Examiner's secrets

From the information given in the table, you must assume a structure for N_2O_4 which is not strictly correct. See page 99.
Niceties of chemistry are sometimes ignored to set you a more straightforward problem.

5 (a) Applying Hess's law to each of the following,
   $C_6H_6 + 7\frac{1}{2}O_2 \rightarrow 6CO_2 + 3H_2O \quad \Delta H_f = +45$ kJ mol^{-1}
   $C_6H_8 + 8O_2 \rightarrow 6CO_2 + 4H_2O \quad \Delta H_f = +35$ kJ mol^{-1}
   $C_6H_{10} + 8\frac{1}{2}O_2 \rightarrow 6CO_2 + 5H_2O \quad \Delta H_f = -42$ kJ mol^{-1}
   $C_6H_{12} + 9O_2 \rightarrow 6CO_2 + 6H_2O \quad \Delta H_f = -160$ kJ mol^{-1}
   (b) (i) $C_6H_6 + 3H_2 \rightarrow C_6H_{12} \quad \Delta H = -205$ kJ mol^{-1}
       (ii) $C_6H_8 + C_6H_{10} + 3H_2 \rightarrow 2C_6H_{12} \quad \Delta H = -313$ kJ mol^{-1}
   In both (i) and (ii) three double bonds have been hydrogenated. Less energy is given out from benzene because the molecule is more stable due to delocalization.

6 (a) $Pt\,|\,Cl^-(aq),\ Cl_2(g)\ \vdots\vdots\ 2H^+(aq)\,|\,[H_2(g)]Pt$
   (b) $1.51 - 1.33 = +0.18$ volts
   (c) Since the standard redox potential of the manganate(VII) system indicates that the oxidation of chloride ions to chlorine is feasible, sulphuric acid is used to produce aqueous hydrogen ions so that no manganate(VI) ions are involved in oxidation of chloride ions.
   (d) There would be no effect, since the total e.m.f. for such a cell would be negative and the reaction would not be feasible.

Index